The Classification and Analysis of Organic Compounds

THE CLASSIFICATION AND ANALYSIS OF ORGANIC COMPOUNDS

Edited by:

Dr. Tirth

ARCLER
PRESS

www.arclerpress.com

The Classification and Analysis of Organic Compounds

Dr. Tirth

Arcler Press
224 Shoreacres Road
Burlington, ON L7L 2H2
Canada
www.arclerpress.com
Email: orders@arclereducation.com

ISBN: 978-1-77469-842-6 (Hardcover)

ABOUT THE EDITOR

Dr. Tirth is presently working as an assistant professor in the department of Chemistry, Faculty of Sciences, Parul University. He did his M.Sc. & Ph.D. in Chemistry from the Maharaja Sayajirao University, Vadodara. His research interest is in Drug Discovery, Synthetic Organic Chemistry and Natural Products isolation. To add to her credits, he has published 15 articles in various international journals. He is having 4 U.S. & 1 Indian patents so far.

TABLE OF CONTENTS

LIST OF FIGURES

LIST OF ABBREVIATIONS

2,4,6-TCP	2,4,6-Trichlorophenol
2,4-DCP	2,4-Dichlorophenol
AO	Atomic Orbital
APIs	Active Pharmaceutical Ingredients
BP	Boiling Points
CASE	Computer-Aided Structural Evaluation
CCD	Charge Coupled Device
CFD	Computational Fluid Dynamics
CNDO	Complete Neglect of Differential Overlap
CPAs	Critical Process Attribautes
CW	Continuous Wave
DCBs	Dichlorobenzenes
DCP	Dichlorophenol
ECD	Electron Capture Detection
EM	Effective Molarity
EPA	Environmental Protection Agency
ETS	Environmental Tobacco Smoke
FBRM	Backscattering Devices
FBRM	Focused Beam Reflectance Measurement
FID	Flame Ionization Detection
FTIR	Fourier Transform Infrared
FWAs	Fluorescent Whitening Agents
GC	Gas Chromatography
GCM	Group Contribution Model
GI	Digestive System
GI	Group Interchange
GIMs	Group Interchange Models
HPLC	High-Performance Liquid Chromatography
IR	Infrared
IUPAC	International Union of Pure and Applied Chemistry
IVR	Intramolecular Vibrational Redistribution
LCAO	Linear Combination of Atomic Orbitals

LCDs	Liquid Crystal Displays
MAO	Monoamine Oxidase
MCB	Monochlorobenzene
MCIs	Molecular Connectivity Indices
MCP	Microchannel Plate
MCP	Mono Chlorophenol
MO	Molecular Orbital
MPs	Melting Points
MW	Microwave
NASA	National Aeronautics and Space Agency
NH_3	Ammonia
NIR	Near Infrared
OLEDs	Organic Light-Emitting Diodes
PATs	Process Analytical Technologies
PCP	Pentachlorophenol
PNMT	Phenylethanolamine-N-Methyltransferase
PSD	Particle Size Distribution
PSP	Pressure-Sensitive Paint
QSPRs	Quantitative Structure-Property Relationships
RDF	Radial Distribution Function
RGB	Red, Green, and Blue Light
SAM	S-Adenosylmethionine
SBS	Sick Building Syndrome
SMILES	Simplified Molecular Input Line Entry System
SMS	Single-Molecule Spectroscopy
SOC	Spin-Orbit Coupling
TCBs	Trichlorobenzene
TCBT	Tetrachlorobenzyltoluenes
TCP	Trichlorophenol
THMs	Trihalomethanes
TSCA	Toxic Substances Control Act
UDPGA	UDP-Glucuronate
UV/Vis	Ultraviolet/Visible
VSEPR	Valence Shell Electron Pair Repulsion

PREFACE

Organic compounds and the reaction products of these chemicals are increasingly observed in biological and physical environments. These compounds can be found and measured on all scales, from the individual to global, in every sector of the environment. While certain biogenic Volatile Organic Compounds or VOCs serve a variety of biophysical and biochemical purposes in living cells and microorganisms, others are released into the environment by living things, sometimes in large amounts. Diverse vegetation, including the prairies of North America and the Central European woodlands, emit isoprene and terpenes into the air, and during the rice-growing season, large amounts of methane are produced and released into the atmosphere from Asian paddy fields. These emissions are caused by marine microorganisms. Although the source of this methane may be biological, its occurrence and quantity are anthropogenically regulated because they depend on the production of rice. Organic compound emissions grew enormously as the Industrial Revolution picked up speed in the 19th century. The chemical industry produced a wide variety of substances in pre-war Germany, including novel VOCs. The delicate ecosystemic balance has been disturbed as a result of the environment and living things being exposed to VOCs, which has also had detrimental health repercussions. According to toxicological research, many VOCs have both reversible and irreversible effects on the human., ranging from short-term effects to long-term effects like the development of cancer. According to ecotoxicological studies, there are repercussions on the environment that vary from the extinction of fragile species to changes in the population of terrestrial and aquatic ecosystems.

Because active pharmaceutical ingredients (APIs) are created in solid form, crystallization is a crucial process in the drug industry. The physical chemistry of organoelemental compounds, which act as a link between inorganic and organic chemistry, is undergoing rapid expansion. The way that matter absorbs light expands the scope of physical and chemical processes. All-natural sciences, as well as a wide range of technical fields, are part of the interdisciplinary topic of photochemistry by nature. This book intends to give a practical guide for students at the undergraduate and graduate levels, as well as for researchers in other fields.

Introduction

The branch of chemistry that examines carbon and its compounds is known as organic chemistry. Today, scientists are aware of an almost infinite variety of compounds that carbon can create. Our daily lives are impacted by the employment of organic chemicals in agriculture, health, and everyday life. According to Oparin, amino acids may have been created during the big bang when ammonia, nitrogen, carbon dioxide, and methane got mixed. This experiment has been confirmed in the lab by Miller. Ancient Romans and Egyptians employed organic chemicals as natural dyes, medicines, and poisons, but it was unknown what the chemical composition was. Organic compounds were separated in their purest form from nature in the 16th century, and analytical techniques for determining elemental composition were framed. We introduce organic chemistry here because covalent compounds make up the majority of organic substances. Conventionally, even though they contain carbon, molecules, including carbonate ions and bicarbonate ions, as well as carbon dioxide and carbon monoxide, are not regarded as belonging to organic chemistry. Hydrocarbons, which are exclusively made of carbon and hydrogen atoms, are the most basic organic molecules. Some hydrocarbons are composed of only single bonds and take the form of a chain of carbon atoms and hydrogen atoms that can be either straight or have branches.

Alkanes are hydrocarbons containing single carbon-carbon bonds, and their names vary based on the number of carbon atoms in the molecule. Carbon-carbon double bonds are present in a different class of substances known as alkenes. Additionally, these double bonds are found in fats, and there is growing evidence linking their inclusion or exclusion in the diet to health problems. Long-chain organic compounds (fatty acids) and glycerol are combined to form fats. At room temperature, saturated fats, such as fat, are often solids. Mono or polyunsaturated fats, which are frequently referred to as oils, are likely to be liquids at room temperature. Mono or polyunsaturated fats include olive oil, flaxseed oil, and several fish oils. Alcohol molecules include an OH (hydroxyl) functional group. Alcoholic beverages contain ethanol, often known as ethyl alcohol. Other alcohols include 2-propanol, which is used as a medical disinfectant, and methanol, commonly used as a solvent. Both methanol and isopropyl alcohol should not be consumed since they are poisonous, even in small amounts. This book explains some of the introductory concepts of organic chemistry.

CHAPTER

1 Introduction to Organic Compounds

Contents

1.1. INTRODUCTION

From the clothes we wear to the pixels on our television and computer screens to the preservatives in food, organic chemistry is present in every part of our lives. We will genuinely have the ability to influence society if we invest the time to comprehend organic chemistry and acquire its mechanisms. Organic chemistry gives scientists the ability to create new drugs, design molecules that can speed up computer processors, and create strategies for reducing the caloric content of foods without compromising their quality (Robinson & Woollard, 1982). It can demonstrate how we can extend and enhance life and explain biological processes like aging, brain functioning, and cardiac arrest. The study of chemical substances containing the element carbon is known as organic chemistry (Figure 1.1).

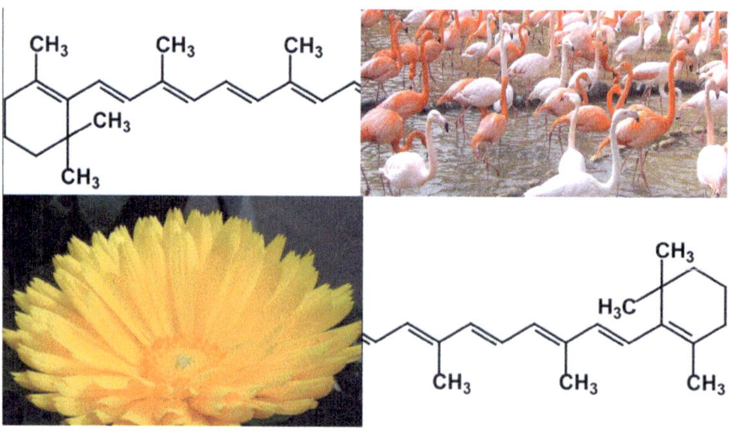

Figure 1.1. *Organic chemistry.*

Source: https://upload.wikimedia.org/wikipedia/commons/1/1f/Front_cover_Organic_Chemistry.jpg, Creative Commons License.

An inorganic compound does not contain the element carbon. There are many causes, but the main one is that carbon compounds are essential to life on Earth since they make up the framework of living things. Carbon-based molecules are what makes us exist (Hoehler, Som, & Kiang, 2018). The ability of carbon atoms to create strong bonds with other carbon atoms to form rings and chains of carbon atoms, as well as the ability of carbon atoms to form strong connections with other elements including hydrogen, nitrogen, oxygen, and sulfur, are the key factors. Carbon serve as the foundation for the vast variety of chemicals required for the formation of life due to its ability to form bonds. It is extremely unlikely that silicon could serve as the building block for anything comparable to life as we know it since the bonds that silicon atoms establish with one another are not nearly as strong as those produced by carbon. The work of physicists and cosmologists has helped us to comprehend a large portion of the origin of the elements (Smeenk & Ellis, 2017). At the beginning of time, the light elements hydrogen and helium were created. The next three elements, lithium, beryllium, and boron were created shortly, after the cosmos had begun to cool. Millions of years later, in the cores of stars, all the heavier elements were created by processes in which the nuclei of lighter atoms fuse to make heavier elements.

The fusion of hydrogen atoms (Figure 1.2) into helium atoms is the main source of energy for stars (McCracken & Stott, 2005). Some stars eventually exhaust their hydrogen supply, collapse, and explode and they are known as supernovae. The explosions of supernovae disperse heavy materials throughout space. Some of these heavy elements were eventually incorporated into the mass of planets like the Earth after being dragged by gravity. Before studying carbon compounds, it is important to examine certain fundamental yet well-known concepts regarding chemical components and their structure. Elements are made up of atoms, and the compounds we encounter in chemistry are composed of elements mixed in various ratios. A compact, positively charged nucleus with protons and neutrons, as well as an outside cloud of electrons, makes up an atom.

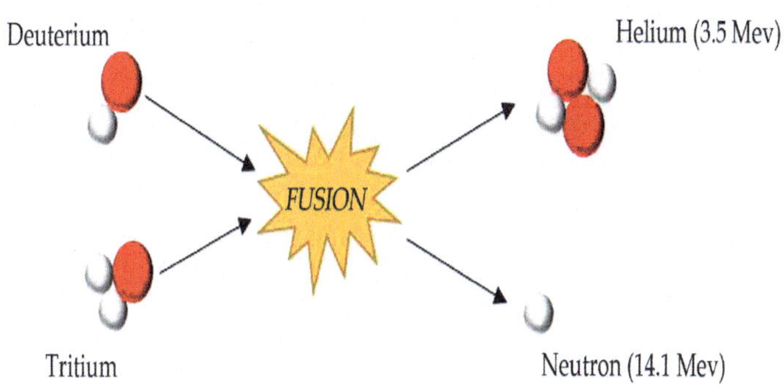

Figure 1.2. Fusion of hydrogen atoms.

Source: https://assets.weforum.org/editor/d2oJijkD2AEUGk_ozW_7nc9nWWWU8JQMte-NOV_vb_0.PNG, Creative Commons License.

Electrons have one negative charge while each proton in the nucleus has one positive charge. Since they have no charge, neutrons are electrically neutral. Protons and neutrons are around 1,800 times heavier than electrons and have nearly similar weights (Grozdanov & Solov'ev, 2020). Therefore, the mass of the nucleus makes up the majority of an atom's mass, with the electrons' contribution to the atomic mass being quite less. However, the electrons account for the majority of an atom's volume; their volume is around 10,000 times greater than that of the nucleus.

1.2. ORGANIC COMPOUNDS

Carbon, hydrogen, nitrogen, oxygen, phosphorus, sulfur, and the halogens like fluorine, chlorine, bromine, and iodine are the elements most frequently found in organic compounds (House & House, 2016). The number of protons in each nucleus, or atomic number (Z), serves as an identifying characteristic for each element. The number of electrons encircling the nucleus is equal to the atomic number since an atom is electrically neutral. For instance, the atomic number of carbons is six due to the presence of six protons in the nucleus. Since each proton and each neutron add one atomic mass unit (1 amu) to the mass of the atom, the majority of carbon atoms also have six neutrons. As a result, ^{12}C is the symbol for these types of carbon atoms, which have a mass number of 12. The same element's atoms will all have the same number of protons in their nucleus, but some of those atoms may have different masses because they have varying numbers of neutrons. Isotopes are the name for these atoms. Around 1% of the atoms of elemental carbon, for instance, have nuclei with 7 neutrons, giving them a mass number of 13. These atoms have the symbol ^{13}C. There are 8 neutrons and a mass number of 14 in the nucleus of a small percentage of carbon atoms. Atoms of carbon-14 (Figure 1.3) are radioactive, as opposed to those of carbon-12 and carbon-13. Using the $_{14}C$ isotope, carbon dating is performed (Fallon, 2011).

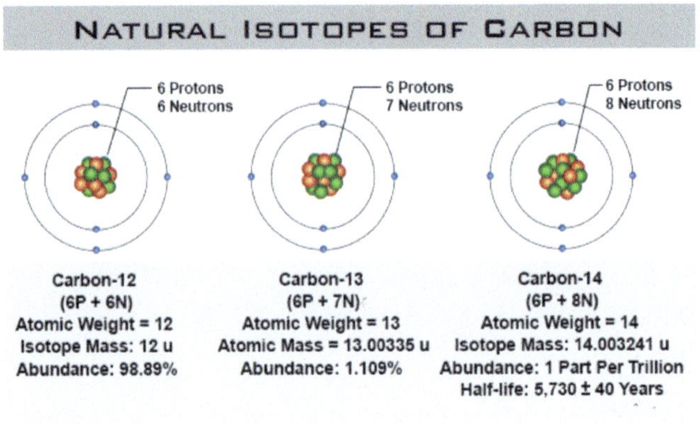

Figure 1.3. Carbon isotopes.

Source: https://open.oregonstate.education/app/uploads/sites/37/2019/07/carbon-isotopes.jpeg, Creative Commons License.

1.3. CHEMICAL BONDS

The majority of hydrogen atoms contain one proton and no neutron in their nuclei. They are represented as ^{1}H and have a mass number of 1. However, just one neutron (0.015%) of the naturally occurring hydrogen atoms is present in the nucleus (Kravchenko, Liberman, & Johansson, 1996). These atoms, known as deuterium atoms, have a mass number of 2. Tritium is an unstable and radioactive isotope of hydrogen with two neutrons in its nucleus. In 1916, G. N. Lewis of the University of California, Berkeley, and W. Kössel of the University of Munich presented the first explanations of the nature of chemical bonds (Kohler, 1975). There were two main categories of chemical bonding put forth: (i)To make ions, one or more electrons from one atom are transferred to another to form ionic or electrovalent bonds.. (ii) Atoms that share electrons form covalent bonds. Their research on bonds is centered around the idea that because noble gas electronic configurations (Figure 1.4) are known to be extremely stable, atoms without them typically react to create them.

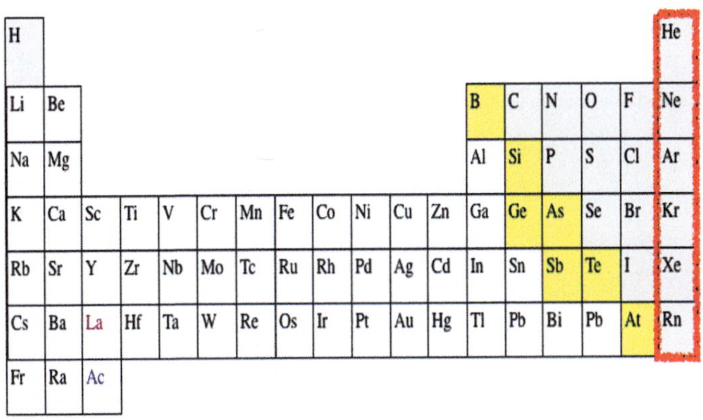

Figure 1.4. *Noble gas electronic configuration.*

Source: http://butane.chem.uiuc.edu/pshapley/GenChem1/L6/ptable.png, Creative Commons License.

To achieve an octet of electrons in the valence shell for all noble gases other than helium. The valence shell is an outermost electron shell, and the octet rule describes an atom's propensity to adopt a configuration in which its valence shell has eight electrons. For many of the issues we currently face in organic chemistry, the ideas and explanations that follow from Lewis and Kössel's original hypotheses are sufficient.

1.4. IONS

Ions are charged particles that are formed when atoms gain or lose electrons. An ionic bond is the attraction of 2 oppositely charged ions. An interaction between atoms with vastly different electronegativities is one source of these ions. The ability of an atom to draw electron is measured by electronegativity. On the periodic table, electronegativity rises as we move up a vertical column and across a horizontal row, respectively. The reaction between lithium and fluorine atoms is an illustration of how an ionic bond is formed.

The lithium atom (Figure 1.5) releases a lithium cation (Li^+), while the fluorine atom gains an electron to release a fluoride anion (F^-). Atoms can acquire the electronic configuration of a noble gas by gaining or losing electrons, which is how ions are created (Miao et al., 2015). With 2 electrons in its valence shell, the lithium cation is equivalent to an atom of the noble gas helium, while the fluoride anion, which has eight electrons, is equivalent to an atom of the noble gas neon. In addition, the lithium and fluoride ions combine to form crystalline lithium fluoride. In this reaction, positive lithium ions are surrounded by negative fluoride ions whereas negative fluoride ions are surrounded by positive lithium ions.

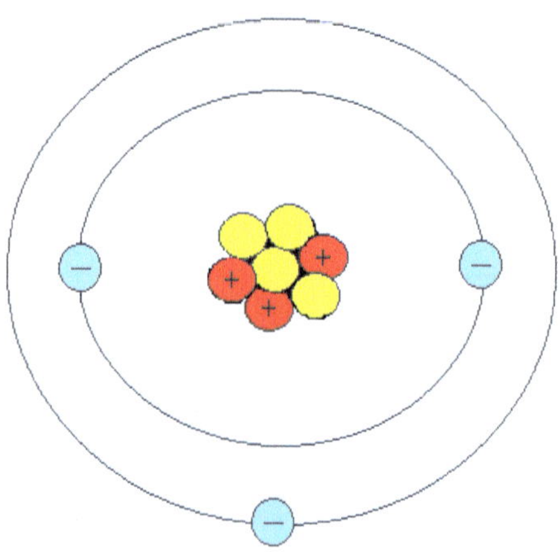

Figure 1.5. Lithium atom.

Source: https://upload.wikimedia.org/wikipedia/commons/d/d8/Configuration_of_a_Lithium_Atom.png, Creative Commons License.

The ions are far less energetic than the atoms they were generated from in this crystalline state. Thus, the reaction that produces crystalline lithium fluoride stabilizes lithium and fluorine. Because of their powerful internal electrostatic forces, ionic compounds are high-melting solids with melting temperatures above 1,000°C (Cai et al., 2021). Ions are dissolved in polar solvents like water, and these solutions typically transmit an electric current.

1.5. ATOMS

Atoms with differing electronegativities cannot form ionic compounds, often known as salts, unless they exchange electrons to become ions. All of the hydrogen atoms that are connected to a specific carbon in a formula are typically written right after the carbon. The hydrogens are normally listed first, followed by all of the atoms that are connected to the carbon. For instance, the bond-line structural formula is used in organic chemistry, and these are known as skeleton formulas. Bond-line formulae make it possible to interpret molecular connectivity and compare one molecular formula to another rapidly because they don't contain all of the symbols that are explicitly presented in dash and condensed structural formulas. Any element in the first two rows of the periodic table has an electron configuration that can be determined using these relative energies. Only a few guidelines/rules must be followed.

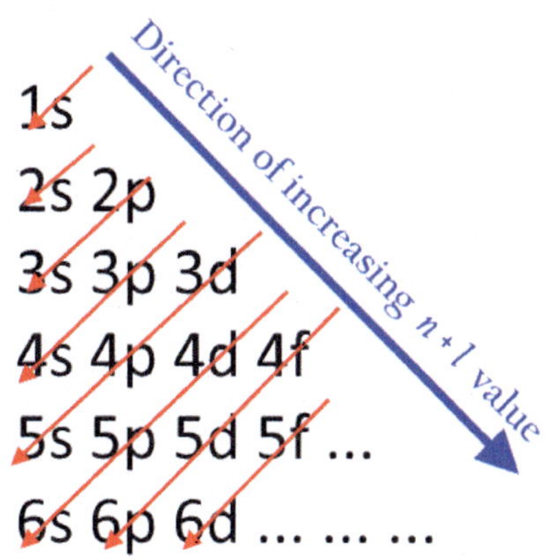

Figure 1.6. Aufbau principle.

Source: https://upload.wikimedia.org/wikipedia/commons/1/15/Aufbau_Principle.png, Creative Commons License.

They are: (i) The Aufbau principle (Figure 1.6) states that the lowest energy orbitals are filled first (Guo et al., 2019). (ii) The Pauli exclusion principle (Figure 1.7) states that each orbital can hold a maximum of two electrons, but only if their spins are coupled (Klyachko, 2009). An electron is only allowed to have one of spin orientations. Arrows are used to depict these orientations. (iii) Hund's rule (Figure 1.8) states that whenever we encounter degenerate orbitals, i.e., orbitals with equal energy, like the three p orbitals, we add one electron to each one while leaving the spins of the electrons unpaired until each degenerate orbital has one electron (van den Brink & Khomskii, 1999). This enables the electrons to be further apart because they repel one another. The spins are then coupled when we start to add an extra electron to each degenerate orbital.

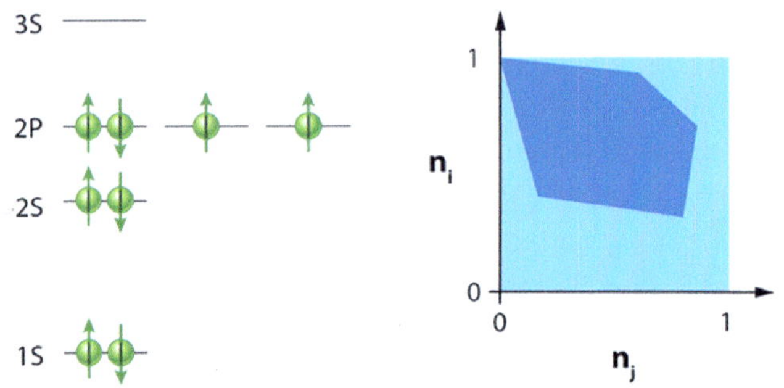

Figure 1.7. Pauli exclusion principle.

Source: https://physics.aps.org/assets/b2d29fef-b9e0-4211-a432-d4103f-f8ace6/e8_1.png, Creative Commons License.

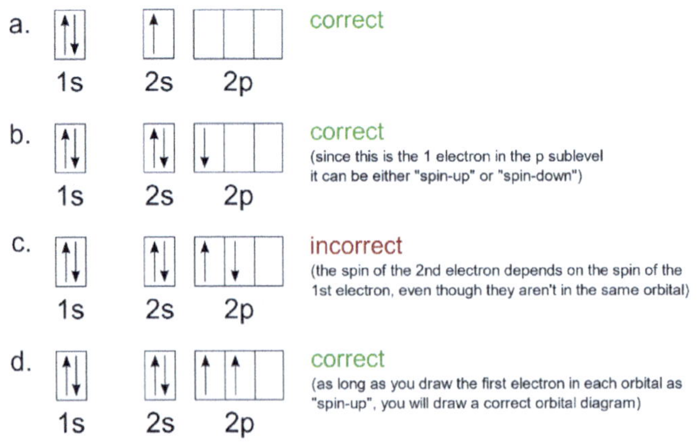

Figure 1.8. Hund's rule.

Source: https://upload.wikimedia.org/wikipedia/commons/8/85/Hund%27s_Rule.svg, Creative Commons License.

Atomic orbitals (AOs) (Figure 1.9) give us a way to comprehend how atoms create covalent bonds. When a covalent bond forms, the system's overall energy is reduced. Each nucleus of the two hydrogen atoms draws the other's electron closer to it as they come closer to one another. The repulsive force that exists between two nuclei is balanced by this attraction. There will be formation of a covalent bond, where the internuclear distance strikes the perfect balance between the sharing of the two electrons between the two atoms and preventing repulsive interactions between their nuclei. The repulsion amid the two positively charged nuclei dominates if the nuclei are pushed closer together, and the system's energy increases.

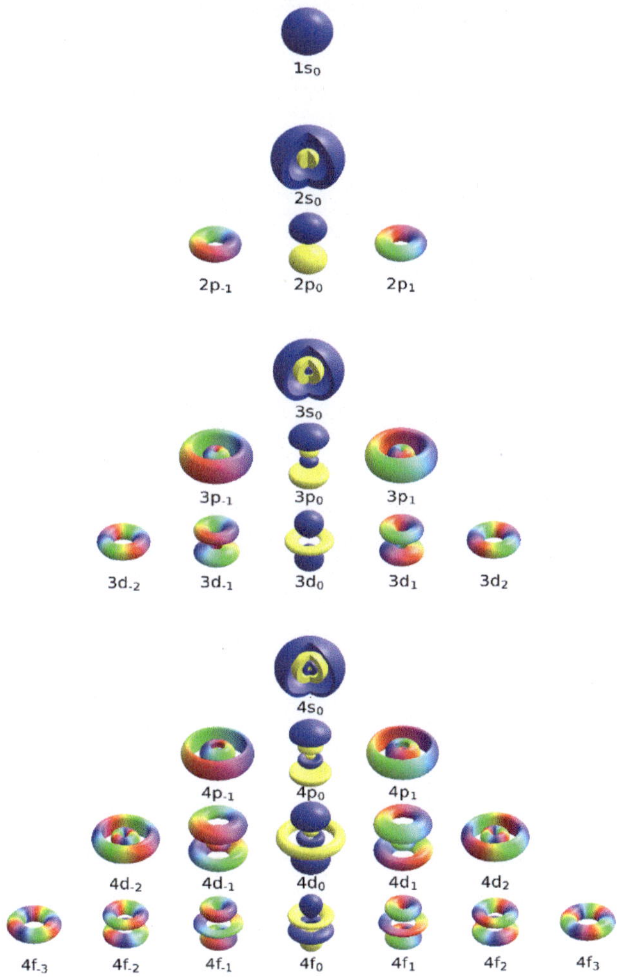

Figure 1.9. Atomic orbitals.

Source: https://upload.wikimedia.org/wikipedia/commons/b/b0/Atomic_orbitals_n1234_m-eigenstates.png, Creative Commons License.

1.6. ORBITALS

The Heisenberg uncertainty principle (Figure 1.10) showed that it is impossible to know an electron's position and momentum at the same time (Busch, Heinonen, & Lahti, 2007). The space where one or two isolated electrons are most likely to be found is called an AO. The two hydrogen atoms' 1s orbital start to overlap as they go closer to one another, eventually joining to create molecular orbitals (MOs).

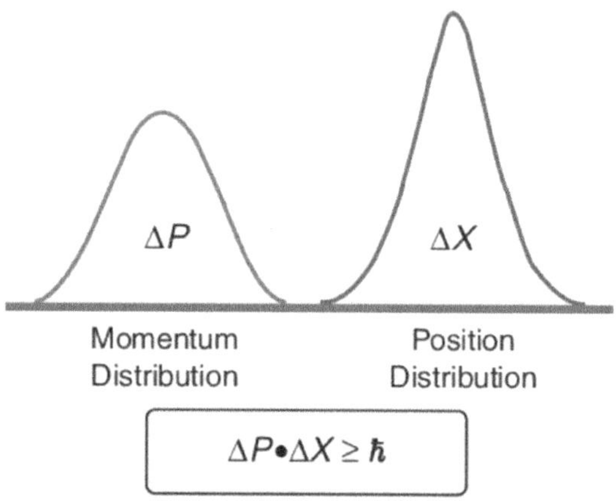

Figure 1.10. Heisenberg uncertainty principle.

Source: https://www.researchgate.net/profile/G_Ali_Mansoori/publication/237677715/figure/fig4/AS:668214166052877@1536326080765/Heisenberg-Uncertainty-Principle_W640.jpg, Creative Commons License.

A MO (Figure 1.11) is the space where one or two of a molecule's electrons are most likely to be located. A maximum of two spin-paired electrons can be found in an orbital (atomic or molecular). The number of MOs that are created when AOs combine to generate MOs is equal to the number of AO or AOs that combine. Thus, two AOs of type 1s combine to generate two MOs during the synthesis of a hydrogen molecule. Because wave functions can be merged mathematically in two ways, i.e., by addition or subtraction. So,They can mix in phase or out of phase. When two orbitals belonging to the same phase overlap, a bonding MO is created. When two orbitals of the same phase overlap, a MO known as an antibonding orbital is created. Both of the electrons from the individual hydrogen atoms are present in the bonded MO of a hydrogen molecule in its lowest energy (ground) state. In the hydrogen molecule's ground state, there are no electrons in the antibonding MO. Since the antibonding orbital does not allow for an electron density between the atoms, bonding is not possible with it.In the LCAO or linear combination of the AOs method, new wave functions for the MOs are created by linearly combining the wave functions for the AOs (either by addition or subtraction) (Froyen & Harrison, 1979).

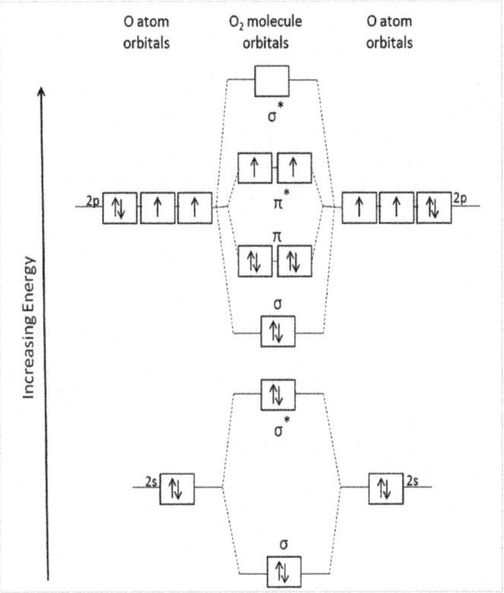

Figure 1.11. Molecular orbital.

Source: https://upload.wikimedia.org/wikipedia/commons/c/ce/Oxygen_molecule_orbitals_diagram.JPG, Creative Commons License.

Similar to AOs, certain electron energy levels correlate to specific MOs. An electron's relative energy in the hydrogen molecule's bonding MO is significantly lower than that of an electron in a s1 AO. These calculations demonstrate that an electron's energy in an antibonding MO is significantly higher than its energy in as 1 AO. Electrons are positioned in MOs the same way they are in AOs. The bonding MO, where two electrons are located, has a lower total energy than the distinct AOs. This is the hydrogen molecule's ground state or lowest electronic state. In what is referred to as an excited state for the molecule, an electron may be present in the antibonding MO. This state is created when a photon of light with the energy (E) is absorbed by a molecule in its ground state. The electrons can be positioned in the hybrid orbitals, always adhering to the Pauli principle that states no more than two electrons may occupy a given orbital. Many of the compounds that we have thought about have four carbon atoms that have used their four valence electrons to create four single covalent (sigma) connections with the other four atoms. However, there are numerous significant organic molecules in which two carbon atoms share more than two electrons. Multiple covalent bonds are generated in some of the bonds in the molecules of these substances. A carbon-carbon double bond is created, for instance, when two carbon atoms share two pairs of electrons. Alkenes are a class of hydrocarbons that have a double bond between two C atoms in their molecules. Alkenes include ethene (C_2H_4) and propene (C_3H_6). Both ethylene and propene are referred to as propylene. The lone carbon-carbon bond in ethene is a double bond. One carbon-carbon single bond and one carbon-carbon double bond are present in propene. Alkene atoms are arranged differently from alkane atoms in terms of spatial arrangement.

1.7. CARBON ATOMS

Ethene has six coplanar atoms with a triangle arrangement of atoms surrounding each carbon atom (Veshapidze et al., 2004). Carbon atoms that have undergone sp^2 hybridization make up carbon-carbon double bonds. The 2s orbital is combined (or hybridized) with two of the 2p orbitals. The electrons are not affected by the hybridization process; only the orbitals are. There is one unhybridized 2p orbital. After that, one electron is added to each of the sp^2 hybrid orbitals while one is left in the 2p orbital. The ethyne carbon-carbon triple bond is shorter than the ethene carbon-carbon double bond, which is also shorter than the ethane carbon-carbon single bond. This is because the hybridization states of the relevant carbon atoms have an impact on bond lengths. The bond is shorter when one or both atoms have a higher s orbital property. This is so because s orbitals are more densely populated with electrons and have a spherical shape than p orbitals. The bond length increases with the p-orbital nature of one or both atoms. This is due to the lobe-shaped and electron density of p orbitals. A sp hybrid orbital, in terms of hybrid orbitals, consists of 50% s character and 50% p character (Bent, 1960). Around 33% of a sp^2 hybrid orbital's character is s, whereas 67% is p (Brewer, 1968). A hybrid sp^3 orbital contains 25% s character and 75% p character (Foster & Weinhold, 2002). Sp hybrid-related bonds are shorter than sp^2 hybrid-related bonds, which are shorter than sp^3 hybrid-related bonds. Both C-C bonds and C-H bonds follow this tendency.

An AO is a region of space around a single atom's nucleus where an electron has a high chance of being present. The s orbital of an atom is spherical, but the p orbital looks like two tangent spheres. When their spins are coupled, orbitals can hold a maximum of two electrons. Wave function is used to characterize orbitals, and each orbital has a unique energy. AOs cross over to create MOs. The space that covers two or more nuclei and contain electrons are known as MOs. If their spins are coupled, MOs, like AOs, can accommodate up to two electrons. A bonding MO is created when two AOs with the same phase sign come into contact. The space between the two nuclei where the negative electrons hold the positive nuclei together has a high electron probability density of a bonding MO. When orbitals with the opposing phase sign overlap, an antibonding MO form. The energy of an antibonding orbital is greater than that of a bonding orbital. The region between the nuclei has a low electron probability density. Electrons in an antibonding orbital do not contribute to the nuclei's ability to remain together.

They frequently go apart as a result of the internuclear repulsions. An electron's energy in a bonding MO is lower than its energy in its individual AO. When two electrons are antibonded, their energy is higher than when they are in distinct AOs. The number of AOs used to create a given number of MOs is always the same. There are always two MOs produced when two AOs are combined, i.e., one bonding and one antibonding. In order to create hybrid AOs, s and p orbitals from the same atom's wave functions are mixed together or hybridized. Four sp^3 orbitals are produced when three p orbitals are hybridized with 1 s orbital. sp^3 hybridized atoms point the axes of their four sp^3 orbitals in the direction of the four corners of a tetrahedron.

1.8. METHANE

Methane's (Figure 1.12) carbon atoms are tetrahedral and sp^3 hybridized (Matxain, Piris, Mercero, Lopez, & Ugalde, 2012). Three sp^2 orbitals result from the hybridization of two p orbitals and 1 s orbital. The axes of the three sp^2 orbitals on sp^2 hybridized atoms point in the direction of the corners of an equilateral triangle. Ethene has sp^2 hybridized, trigonal planar carbon atoms. Two sp orbitals result from the hybridization of one p orbital with 1 s orbital. sp-hybridized atoms have their two sp orbital axes pointed in opposite directions at an angle of 180°. Ethyne is a linear molecule and has sp hybridized carbon atoms. A sigma bond is a form of single bond in which, when examined along the bond axis, the electron density exhibits circular symmetry. The skeletons of organic molecules are typically made up of atoms connected by sigma bonds. A pi bond is a type of double or triple carbon-carbon bond in which the sideways overlapping electron densities of two parallel neighboring p orbitals produce a bonding pi MO. The valence shell electron pair repulsion (VSEPR) model (Figure 1.13) is a concept that allows us to predict how atoms will be arranged in molecules and ions (Gillespie, 1963). Molecules (or ions) whose core atom is covalently connected to two or more other atoms or groups can be considered.

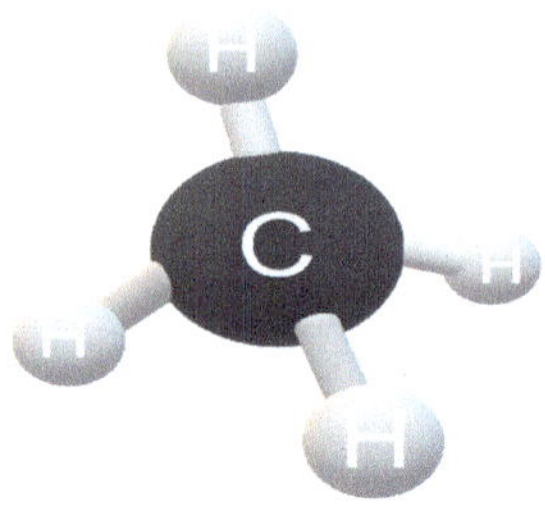

Figure 1.12. Methane.

Source: https://upload.wikimedia.org/wikipedia/commons/9/94/Methane_CH4.png, Creative Commons License.

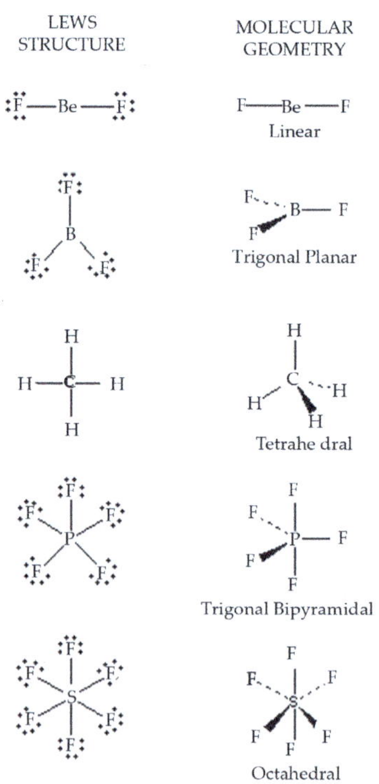

Figure 1.13. *Valence shell electron pair repulsion (VSEPR) model.*

Source: https://chemed.chem.purdue.edu/genchem/topicreview/bp/ch8/graphics/fig8_20.gif, Creative Commons License.

All of the central atom's valence electron pairs are taken into account, including those that are shared in covalent bonds and are referred to as bonding pairs as well as those that are not, which are referred to as nonbonding pairs, unshared pairs, or lone pairs. The electron pairs of the valence shell keep as much space between them as possible since they oppose one another. Nonbonding pair repulsion is stronger than bonded pair repulsion. We need to consider all of the electron pairs, both bonding and nonbonding to determine the geometry of the molecule, but we define the form of the molecule or ion by referring to the positions of the nuclei (or atoms), not the positions of the electron pairs. Methane has four pairs of bonding electrons in its valence shell. Four pairs of electrons can only be separated from one another equally and in a tetrahedral arrangement. Any other orientation brings some electron pairs closer together than others, such as a square planar configuration. Methane has a tetrahedral form as a result.

1.9. CONFIGURATION

The tetrahedral configuration of an ammonia electron pair emerges when the non-bonding electron pair occupies one corner (Figure 1.14). This arrangement of electron pairs explains why the NH_3 molecule has a trigonal pyramidal structure. Unshared electrons are not visible in ball and stick models. A trigonal pyramid is the form of an ammonia (NH_3) molecule. One nonbonding pair and three bonding pairs of electrons are present. Ammonia molecules have bond angles of 107°, which is near to the tetrahedral angle (109.5°) (Gillespie, 2002). The trigonal pyramidal configuration of the four atoms is explained by a tetrahedral arrangement of the electron pairs. Because the nonbonding pair takes up more area than the bonding pairs, the bond angles are 107° (and not 109.5°). The form of a water molecule is angular or twisted. A water molecule's H-O-H bond angle is 104.5°, which is also near to the 109.5° bond angles of methane (Chidambaram, 1962). If we rearrange the two bonding electron pairs and the two nonbonding electron pairs at the corners of the tetrahedron, we may describe a general tetrahedral structure for the electron pairs of a water molecule. The three atoms' angular arrangement is explained by the tetrahedral arrangement. Because the nonbonding pairs are in fact bigger than the bonding pairs and the structure is not exactly tetrahedral, the bond angle is less than 109.5°. There are just three valence electrons in the group IIIA element boron. These three electrons are shared by three fluorine atoms in the molecule boron trifluoride (BF_3) (Figure 1.15).

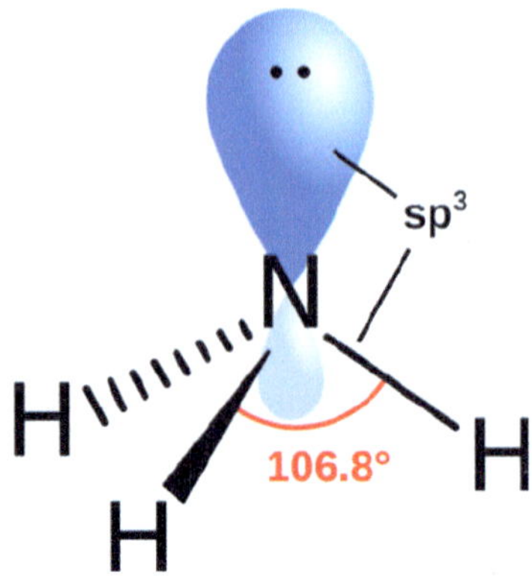

Figure 1.14. Ammonia.

Source: https://www.researchgate.net/publication/230864138/figure/fig1/AS:300560575156225@1448670633553/D-structure-of-ammonia-as-an-example-of-an-N-atom-with-a-tetrahedron-shaped-bond.png, Creative Commons License.

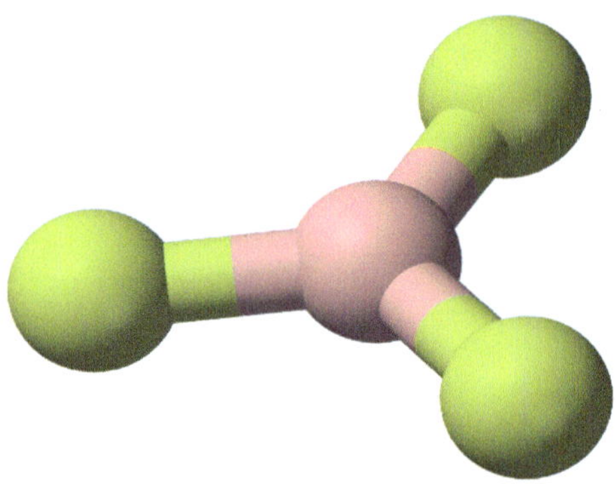

Figure 1.15. Boron trifluoride.

Source: https://upload.wikimedia.org/wikipedia/commons/0/01/Boron-trifluoride-3D-balls.png, Creative Commons License.

As a result, there are only six electrons (three bonding pairs) surrounding the boron atom in BF_3. When three bonding pairs are at the corners of an equilateral triangle, their maximum separation occurs. As a result, the three fluorine atoms in the boron trifluoride molecule are located at the corners of an equilateral triangle in a plane. Only two electron pairs, both of which are bonding pairs, surround the core beryllium atom in BeH_2 (Montero-Campillo, Lamsabhi, Mó, & Yáñez, 2013).

If we assume that all of the electrons of a multiple bond act as though they were a single unit and are therefore situated in the region of space between the two atoms united by multiple bonds, the VSEPR approach can also be used to predict the morphologies of molecules with multiple bonds. The design of a carbon dioxide molecule serves as an example of this concept. Each oxygen atom has a double bond with the core carbon atom of carbon dioxide. The bond angle of carbon dioxide is 180 degrees, and it is known to have a linear shape. The two groups of four bonding electrons can be separated from one another by a maximum of this arrangement. The shape is unaffected by the nonbonding pairs connected to the oxygen atoms. Organisms produce organic molecules across the planet that are made of carbon, hydrogen, nitrogen, and oxygen. These substances help organisms function normally on a global scale and/or help them survive against predators. Numerous distinct compounds with a wide range of properties make up organic molecules. For instance, citrus vitamin C guards against oxidative stress while the chlorophyll in green plants uses sunlight to its advantage. Other compounds include capsaicin, (Figure 1.16) a substance produced by pepper plants that deter insects and birds from attempting to consume them. They consist of lovastatin, a substance found in oyster mushrooms that defends the mushroom against bacterial infections, and salicylic acid, a signaling hormone produced by willow plants.

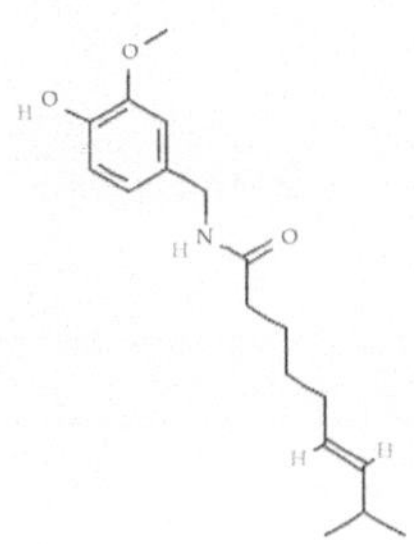

Figure 1.16. Capsaicin.

Source: https://pubchem.ncbi.nlm.nih.gov/image/imgsrv.fcgi?cid=1548943 &t=l, Creative Commons License.

1.10. MEDICAL DRUGS

Capsaicin works well as an analgesic (O'Neill et al., 2012). It is currently offered under the brand name Capzacin and can modulate pain when administered topically. Salicylic acid is used to treat pain as well as acne, while lovastatin is a treatment that lowers blood cholesterol levels in people. Modern organic chemistry's strength is in its capacity to produce such molecules, which are occasionally found in very small amounts in nature, on a massive scale from easily accessible and reasonably priced materials so that everyone in society can take advantage of them. Though some people might believe that natural vitamin C is healthier, the synthetic compound is equally effective because they are chemically identical. For instance, even though we can get vitamin C from eating certain fruits, chemists can make large quantities in the lab for use in daily supplements. Organic chemistry offers the chance to alter the structures of these and other natural productto to create molecules with unique and possibly even more remarkable capabilities. For instance, aspirin, a molecule with much more efficacy as a painkiller and fewer side effects than natural substances, was discovered by adding a few atoms to salicylic acid through a chemical reaction. In reality, slightly more than half of the top 20 medications by gross sales are natural products or their derivatives (Calixto, 2019). Hydrocarbons are substances that exclusively have carbon and hydrogen atoms. The hydrocarbons methane (CH_4) and ethane (C_2H_6) are two examples.

1.11. NOMENCLATURE

Double or triple bonds between the C atoms of other hydrocarbons are possible. An -ene ending in the family name and compound names of alkenes denotes the presence of at least one carbon-carbon double bond (Figure 1.17). Aromatic compounds have a ring, the most dominant example of which is a benzene ring. Alkynes possess at least one carbon-carbon triple bond, and this is reflected in the family name and names for specific compounds by a -yne ending or suffix. In general, compounds like alkanes, whose molecules solely have single bonds, are referred to as saturated compounds. Alkenes, alkynes, and aromatic hydrocarbons are examples of molecules having more bonds. These compounds are known as unsaturated compounds because they have fewer hydrogen atoms than the maximum allowed and can react with hydrogen. Alkanes mostly come from petroleum and natural gas. Under normal circumstances, the smaller alkanes (methane through butane) are gases. Natural gas has methane. Petroleum refining is a major method for obtaining higher molecular weight alkanes. The simplest alkane, methane, was a significant part of this planet's early atmosphere. The atmosphere of the Earth still contains some methane, but not in significant quantities. However, it has a significant role in the composition of the atmospheres of Jupiter, Saturn, Uranus, and Neptune. Methane is created by some living things using carbon dioxide and hydrogen. These extremely basic species, known as methanogens, may be the planet's oldest living things and constitute a distinct evolutionary stage. Only an anaerobic (i.e., oxygen-free) environment is able to support methanogens. They have been discovered in sewage, cow stomachs, and ocean trenches.

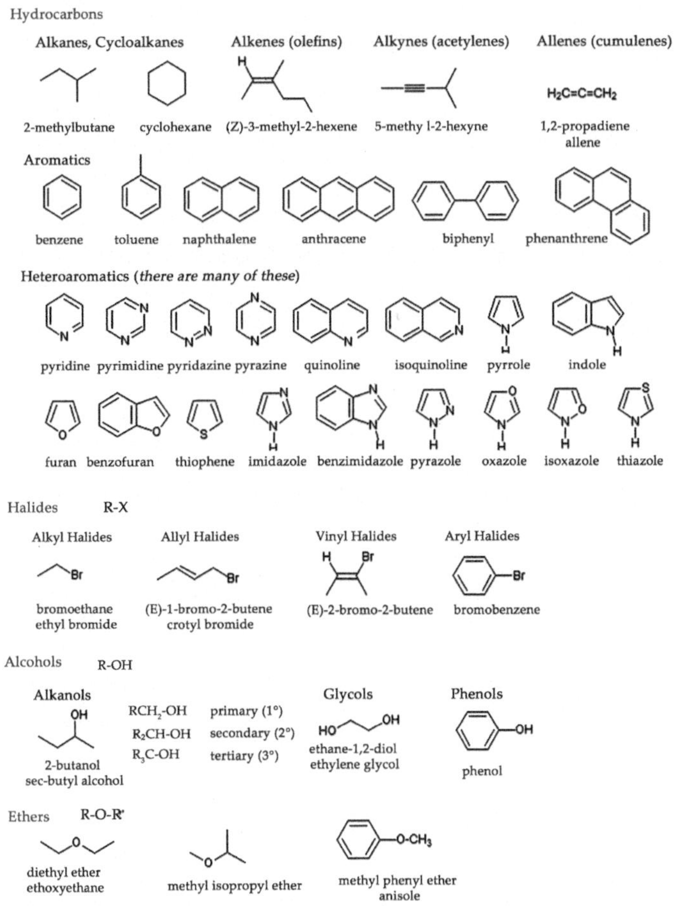

Figure 1.17. Nomenclature.

Source: https://res.cloudinary.com/ocd/image/fetch/q_auto,f_auto/https://organicchemistrydata.org/hansreich/resources/nomenclature/nomenclature_data/func-groups-00.gif, Creative Commons License.

The two simplest alkenes, ethene, and propene, are among the industrial chemicals made in the United States. Around 30 billion pounds of ethene and almost 15 billion pounds of propene are produced annually by the chemical industry (Ittel, Johnson, & Brookhart, 2000). Many industrial chemicals, such as ethanol, ethylene oxide, ethanal, and the polymer polyethylene, are synthesized from ethene. In addition to its many other uses, propene is the starting ingredient for the synthesis of acetone and cumene. Propene is used to create the polymer polypropylene. Ethene also exists naturally in plants as a hormone. It is naturally made from fruits, like tomatoes and bananas, and it helps these fruits mature. Since green fruits are less likely to be harmed during shipping, ethene is frequently used in the commercial fruit sector to hasten the ripening of tomatoes and bananas that have been picked green. Ethyne is the most basic alkyne (also called acetylene). Alkynes are found in nature and can be created in a lab. Capillin, an antifungal drug, and dactylyne, a marine natural product that inhibits pentobarbital metabolism, are two examples of alkynes among many that have a biosynthetic origin. A synthetic alkyne with estrogen-like characteristics, ethinyl estradiol is used in oral contraceptives. The archetypal aromatic chemical is the substance known as benzene. The Kekulé structure of benzene, named after August Kekulé, is a six-membered ring with alternate single and double bonds (Rocke, 2015).

1.12. STRUCTURE

Despite the fact that the Kekulé structure (Figure 1.18) is often employed for benzene compounds, there is more data to suggest that it is insufficient and inaccurate. In actuality, the length of each carbon-carbon bond in benzene is the same, which falls between the ranges of a single and double bond. We understand that benzene cannot be fully represented by either structure or must instead be viewed as a hybrid of the two structures based on the concepts of resonance theory. This hybrid is shown as a hexagon with a circle in the center. The difficulty we had in comprehending why all of the C-C bonds are the same length is thus resolved by resonance theory. The resonance hypothesis claims that the bonds are actually a combination of single and double bonds rather than alternate single and double bonds. Each bond that is a single bond becomes a double bond, and vice versa. In truth, all of the carbon-carbon bonds in benzene are one and a half bonds long, which is halfway between a single bond and a double bond. The C atoms in the benzene ring are sp^2 hybridized and have bond angles of 120° in the

MO. Each carbon has a p orbital with one lobe above and one lobe below the plane of the ring. The lobes of the p orbitals on the atoms on each side of the ring overlap with the lobes of each p orbital above and below the ring. The computed MO illustrates how this type of p orbital overlap results in a collection of bonding MOs that cover the entire ring's carbon atoms. As a result, the six electrons from these p orbitals, i.e., one from each orbital are delocalized around the ring's six carbon atoms.

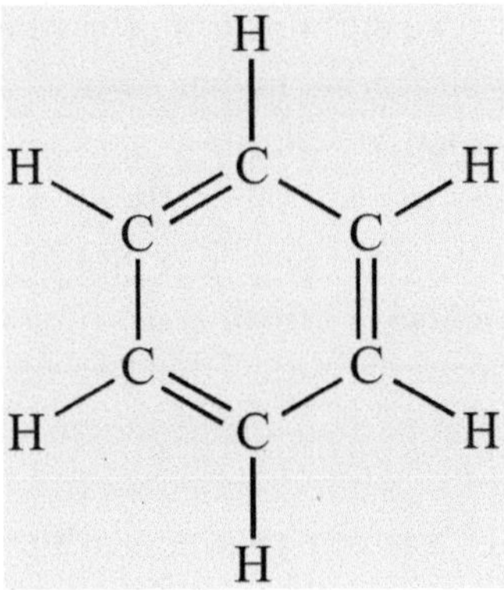

Figure 1.18. *Kekulé structure.*

Source: http://www.chem.ucla.edu/~harding/IGOC/K/kekule_structure01.png, Creative Commons License.

The equal length and equivalence of each carbon-carbon bond are explained by the delocalization of electrons. Polar covalent bonds have a significant impact on the reactivity and physical characteristics of molecules. These polar covalent bonds frequently make up functional groups. In a molecule, functional groups are atoms that contribute to the function. Atoms with variable electronegativity values and unshared electron pairs are frequently found in functional groups. Heteroatoms are described as having unshared electron pairs and forming covalent bonds. Examples include oxygen, nitrogen, and sulfur. The distribution of charge within a molecule can be seen by using an electrostatic potential map. The areas of an electron density surface that are more negatively charged than others are highlighted in red. A species that is positively charged would be drawn to these areas. Blue areas indicate areas with lower or higher negative values. Blue areas will draw electrons from another molecule. The color spectrum from red to blue depicts the direction of influence, going from most negative to least negative. As predicted based on the difference in their electronegativity values, it is evident that the positive charge is localized around the hydrogen atom and that the negative charge is concentrated close to the chlorine atom. This also reveals the general shape of the molecule because it is plotted at the low-electron-density surface of the molecule. There must always be a dipole moment in any diatomic molecule if the two atoms are distinct.

Each of the carbon-chlorine bonds in CCl_4 is polar because chlorine has a higher electronegativity than carbon (Liu, Arnold, Betterton, & Smotkin, 2001). The partial negative charge of each chlorine atom contrasts with the strongly positive charge of the carbon atom. However, a carbon tetrachloride molecule has no net dipole moment because it is tetrahedral, where the centers of positive and negative charge coincide. The effects of the bond moments cancel since they are tetrahedrally organized vectors of identical size. The net dipole moment of the chloromethane (CH_3Cl) molecule is 1.87D. Three C-H bonds barely contribute to the net dipole because carbon and hydrogen have electronegativities that are almost identical. Although there is a significant difference in electronegativity between carbon and chlorine, the highly polar C-Cl bond accounts for the majority of the dipole moment of CH_3Cl. Alkenes' cis-trans isomers differ in terms of their physical characteristics. In addition to having differing melting and boiling points (BPs), cis-trans isomers frequently have dipole moments that are noticeably different from one another. Functional groups are atomic configurations that give a molecule predicted reactivity and characteristics. An alkene's carbon-carbon double bond, for instance, is its functional group. The majority of the chemical reactions involving alkenes are those involving the carbon-carbon double bond. An alkyne's triple bond made of carbon and carbon is its functional group. There is no functional group for alkanes. They have carbon-carbon single bonds and carbon-hydrogen bonds in their molecules, although practically all organic compounds also contain these bonds. C-C and C-H bonds are often less reactive than common functional groups.

Propane may yield two alkyl groups, but methane and ethane can only yield one. The propyl group is produced when a hydrogen atom from one of the end carbon atoms is removed, while the isopropyl group is produced when a hydrogen atom from the middle carbon atom is removed. Alkyl halides are substances in which an alkane's hydrogen atom is exchanged out for an atom of a halogen, such as fluorine, chlorine, bromine, or iodine. Alkyl halides include for example, CH_3Cl and CH_3CH_2Br. Haloalkanes is another name for alkyl halides. An alkyl halide has the general formula RX, where X represents fluorine, chlorine, bromine, or iodine. Depending on their degree, alkyl halides can be primary (1°), secondary (2°) or tertiary (3°). The halogen's direct attachment to the carbon atom is the basis for this classification. A carbon atom is referred to as a primary carbon atom and an alkyl halide is categorized as a primary alkyl halide if it is directly connected to just one other carbon while carrying the halogen. The C is a secondary C, and the R-halide is a secondary R-halide if the carbon bearing the halogen is itself directly connected to two additional carbon atoms. The carbon that contains the halogen is a tertiary carbon and the alkyl halide is a tertiary alkyl halide if it is directly connected to three additional carbon atoms.

1.13. HALOGEN

A halogen atom is joined to an alkene carbon to form an alkenyl halide. Such substances have occasionally been referred to as vinyl halides. A molecule known as an aryl halide has a halogen atom attached to an aromatic ring, such as a benzene ring. Methyl alcohol, often known as methanol, is the most basic substance in the family of chemical molecules known as alcohols and has the structural formula CH_3OH. This functional group is a hydroxyl (OH) group bonded to a carbon atom that has undergone sp^3 hybridization. Ethyl alcohol, often known as CH_3CH_2OH, is another kind of alcohol. In terms of structure, alcohols can be seen as either hydroxyl derivatives of alkanes or alkyl derivatives of water. Alcohols are divided into three categories, similar to how alkyl halides are. They are primary (1°), secondary (2°), and tertiary (3°). This classification is based on how much of the carbon the hydroxyl group is directly connected to has been substituted. The carbon is referred to as a primary carbon and the alcohol termed a primary alcohol if there is only one other carbon connected to it. The alcohol is referred to as a secondary alcohol if the carbon atom carrying the hydroxyl group additionally has two other carbon atoms connected to it. The typical formula for ethers is R-O-R or R-O-R′, where R′ may be an alkyl or phenyl group other than R. Ethers can be viewed as derivatives of water in which alkyl groups have taken the place of both hydrogen atoms. Nitrous oxide (N_2O), often known as laughing gas, was originally used as an anesthetic in 1799 and is still used today even though when administered alone it does not generate deep anesthesia (Gillman & Lichtigfeld, 1994). The binding angle at the oxygen atom of an ether is slightly bigger than that of water. Diethyl ether was the first ether to be used to create deep anesthesia in 1842. Since then, a variety of ethers, typically containing halogen substituents, have taken the place of diethyl ether as popular anesthetics. One explanation is that the halogenated ethers are not very flammable, unlike diethyl ether. Desflurane and sevoflurane are two halogenated ethers that are currently utilized for inhalation anesthesia.

1.14. AMINE

Amines can be thought of as organic derivatives of ammonia, just as alcohols and ethers can be thought of as organic derivatives of water. Based on how many organic groups are joined to the nitrogen atom, this classification is made. A main amine is the potent and hazardous stimulant amphetamine. A major amine is dopamine, a significant neurotransmitter whose depletion is linked to Parkinson's disease. Secondary and aromatic amine groups are present in nicotine, the addictive toxin present in tobacco. The carbonyl group, which is formed when a C atom forms a double bond with oxygen, can be found in both aldehydes and ketones. An aldehyde's carbonyl group is linked to one H atom and one C atom except for HCHO, which is the only aldehyde bearing two hydrogen atoms. A carbonyl group that is joined to an oxygen or nitrogen atom can be found in carboxylic acids, esters, and amides. Ants create a stinging substance called formic acid. CH_3COOH, the compound that gives vinegar its sour flavor, is created when specific bacteria interact with wine's ethyl alcohol and induce oxygen to oxidize it. The formulae for amides are $RCONH_2$, RCONHR′, or RCONR′R″, where a carbonyl group is joined to a nitrogen atom carrying a hydrogen and/or an alkyl group. The formula for a nitrile is RCN:(or RCN). A nitrile's carbon and nitrogen undergo sp hybridization. Acyclic nitriles are designated using the -nitrile suffix after the name of the appropriate hydrocarbon in the IUPAC systematic nomenclature. The number one carbon atom belongs to the C-N group. Both acrylonitrile and acetonitrile are accepted common names for CH_3CN and CH_2=CHCN, respectively. An example of a typical sugar is glucose. In addition to having multiple alcohol hydroxyl groups (OH), glucose also has an aldehyde group in one of its forms. Proteins have amide groups, while fats and oils have ester groups.

One of the initial findings in any experiment would undoubtedly be whether a given substance is a solid, a liquid, or a gas. Melting points (MPs) and BPs, which mark the temperatures at which phases change, are among the more easily studied physical parameters. Identification and isolation of organic compounds can also benefit from knowledge of melting and boiling temperatures. If we are aware of the BPs of the by-products and solvents that could be present in the reaction mixture, as well as the BP of the intended product, we can determine whether or not simple distillation will be a practical method for separating our product. Another possible variation of our product is solid. Making accurate predictions of a compound's melting point, BP, and solubilities is necessary if the compound is to be successfully isolated. According to the structure and the forces that exist between molecules and ions, estimates of these macroscopic physical qualities are made. The strength of these intermolecular forces can be determined by the temperatures at which phase shifts take place. The temperature at which the crystalline state and its liquid state are at equilibrium is known as the melting point of that substance.

The strong electrostatic lattice forces between the positive and negative ions in the crystalline structure operate as the ion-ion forces that hold the ions together in the crystalline state if the material is an ionic compound. Each positively charged sodium ion is bounded by negatively charged acetate ions, and vice versa. The organized crystal structure must be broken up into the open structure of a liquid using thermal energy. As a result, sodium acetate melts at a rather high temperature of 324°C (Duruz, Michels, & Ubbelohde, 1997). Ionic molecules have much higher boiling temperatures, which means that most ionic organic compounds break down before boiling. The fact that even totally nonpolar molecules can exist in liquid and solid forms is explained by the forces that act between molecules, which however are weaker than those between ions. Van der Waals forces, which refer to all of these intermolecular forces, are all electrical in nature. Due to the highly polarized carbonyl group, molecules like acetone and acetaldehyde are examples of those with dipoles. Dipole-dipole interactions in the liquid or solid-state cause molecules to align themselves so that the positive end of one molecule faces the negative end of another.

Although weaker than regular covalent bonds, hydrogen bonds are substantially stronger than the dipole-dipole interactions in acetone. Hydrogen bonds have bond dissociation energies of roughly 4–38 kJ/mol. Even though all chemicals have comparable molecular weights, H bonding explains why water, ammonia, and hydrogen fluoride have significantly higher boiling temperatures than methane. At 25°C, water becomes a liquid instead of a gas, which is one of the effects of hydrogen bonding. Water would have a BP close to 80°C without hydrogen bonding, making it a gas at ambient temperature. It is extremely unlikely that life as we know it could have evolved on Earth if this had been the case. The base pairs of double-stranded DNA are joined by hydrogen bonds. Adenine and thymine form hydrogen bond. Guanine and cytosine form hydrogen bonds. Even though the two substances have the same molecular weight, hydrogen bonding is the reason why ethyl alcohol has a substantially higher BP (78.5°C) than dimethyl ether (24.9°C). Ethyl alcohol molecules can establish strong hydrogen bonds with one another because each of their hydrogen atoms is covalently attached to an oxygen atom. Dimethyl ether molecules cannot form strong hydrogen bonds with one another because they lack a hydrogen atom bound to a strongly electronegative atom. Dipole-dipole interactions are weaker intermolecular forces in dimethyl ether. Phenols have greater boiling temperatures than other aromatic hydrocarbons of the same molecular weight because they can also establish potent intermolecular hydrogen bonds. Despite having the same molecular weight, toluene (Figure 1.19) (methylbenzene, bp 110.6°C) and phenol (hydroxybenzene, bp 182°C) have very different BPs.

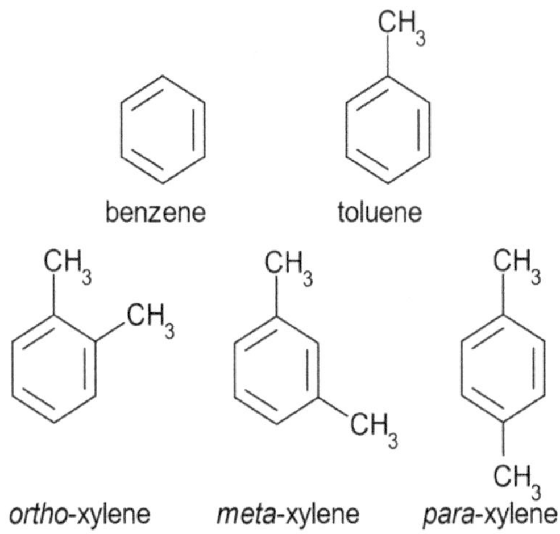

Figure 1.19. *Toluene.*

Source: https://upload.wikimedia.org/wikipedia/commons/thumb/1/11/Benzene_Toluene_and_ortho-%2Cmeta-%2Cand_para-xylene.svg/1200px-Benzene_Toluene_and_ortho-%2Cmeta-%2Cand_para-xylene.svg.png, Creative Commons License.

1.15. QUANTUM MECHANICS

Quantum mechanics (Figure 1.20) must be used for accurately describing the nature of dispersion forces (James, John, Melekamburath, Rajeevan, & Swathi, 2022). In a nonpolar molecule like methane, the average distribution of charge over time is uniform. However, because electrons travel, the electrons and hence the charge might not be evenly distributed at any particular time. An instantaneous minor accumulation of electrons on one region of the molecule may result in the formation of a transient dipole. Because the electron cloud of a nearby section of another molecule will be distorted by the –ve (or +ve) charge in a portion of one molecule, the opposite charge will develop there. Although these transient dipoles are continually shifting, they ultimately serve to create attractive forces between nonpolar molecules, enabling the existence of their liquid and solid phases. The overall pull between molecules brought on by dispersion forces increases with surface area. In comparison to more spherical molecules, molecules that are typically longer, or cylindrical have more surface area accessible for intermolecular interactions. As a result, these molecules exhibit stronger attractive forces than branched molecules. This is clear when contrasting neopentane, the most heavily branched C_5H_{12} isomer in which one carbon bears four methyl groups, with pentane, the unbranched C_5H_{12} hydrocarbon. The boiling point of pentane is 36.1°C. The boiling point of neopentane is 9.5°C. Since their boiling temperatures differ, it can be concluded that pentane molecules are more attracted to one another than neopentane molecules. How polarizable are the electrons in the atoms in question? The ease with which the electrons react to a shifting electric field is referred to as polarizability. Iodine, a large atom, has loosely bound electrons that are easily polarized, whereas fluorine, a small atom, has tightly bound electrons that are far less polarizable.

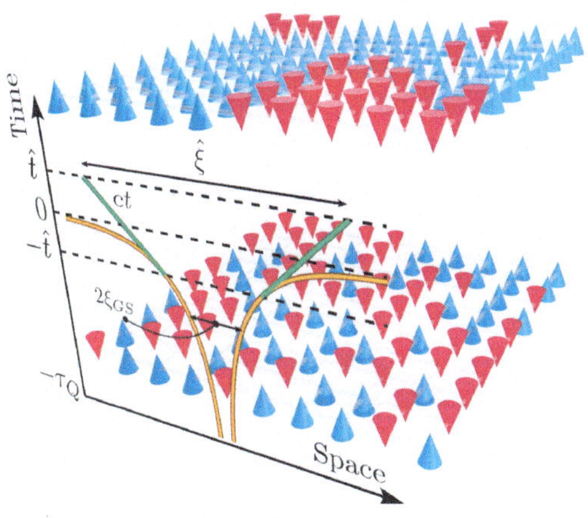

Figure 1.20. *Quantum mechanics.*

Source: https://scx2.b-cdn.net/gfx/news/hires/2022/researchers-answer-fun.jpg, Creative Commons License.

These rapidly varying dispersion forces can add up for large molecules to provide a net attraction. When compared to hydrocarbons with the same molecular weight, fluorocarbons have low boiling temperatures. Despite having a larger molecular weight than pentane (C_5H_{12}), the fluorocarbon C_5F_{12} has a slightly lower BP. The low polarizability of fluorine atoms, which results in minimal dispersion forces, is a key component in explaining this behavior. Making non-stick frying pans and lightweight bearings takes advantage of the self-lubricating qualities of the fluorocarbon known as Teflon $[CF_2CF_2]_n$. The temperature at which a vapor pressure equals the atmospheric pressure above it is known as the BP of the liquid. BPs are always given as a measure at a specific pressure, such as at 1 atm since the BPs of liquids are pressure-dependent. A liquid's BP at 1 atm is considered to be its typical BP. A substance's constituent molecules must separate in order for it to transition from a liquid to a gaseous state. Nonpolar compounds have low boiling temperatures and very weak intermolecular interactions, which make them easily detachable from one another. Bigger molecules have larger surface areas, which can result in stronger intermolecular forces. Heavier molecules need thermal energy to reach adequate velocities to exit the liquid phase. Neopentane's branching form allows for less surface interaction between molecules, which results in lower dispersion forces than pentane's linear structure.

The only forces binding the molecules of pentane together are dispersion forces because they lack polar groups. Its BP would be the lowest. Diethyl ether has a higher BP than pentane because it has the polar ether group, which produces dipole-dipole forces that are stronger than dispersion forces. Since the OH group in sec-butyl alcohol can create powerful hydrogen bonds, it would have a high BP. The primary factor in explaining solubilities is intermolecular forces. Melting a solid is similar to how a solid dissolves in a liquid. As a result of the disturbance of the solid's crystal structure, the molecules (or ions) in the solution are arranged in a more unpredictable manner. The molecules or ions must also be separated from one another during the dissolving process, and both modifications need energy. In order to overcome lattice energies and intermolecular or interionic attractions, new attractive forces between the solute and solvent must first emerge. Lattice energy as well as internal ionic attractions are strong here. The only polar solvents capable of dissolving ionic compounds are water and a few others. These solvents hydrate or solvate the ions to dissolve ionic substances. Due to their strong polarity and small, compact structure, water molecules can efficiently surround the individual ions as they are released from the crystal surface. Negative ions are solvated in the exact opposite manner as positive ions, with water molecules surrounding them with the water dipole's negative end facing the positive ion. As a result of water's high degree of polarity and its ability to form potent hydrogen bonds, the ion-dipole forces of attraction are likewise quite strong. Both the lattice energy and the crystal's internal ionic attractions can be neutralized by the energy provided by the development of these forces.

The hydrophilic ionic groups at the chain ends are exposed on the surface, creating a surface where water molecules are drawn to. Water and oil don't mix, but the oily coating resembles an ionic substance, making water able to clean it. Complex protein molecules fold precisely as a result of the creation of hydrogen bonds, the hydration of polar groups, and the propensity of nonpolar groups to avoid a polar environment, enabling them to act as biological catalysts. The same elements enable hemoglobin molecules to take on the form required for oxygen delivery. They enable lipid-containing proteins and other compounds to serve as cellular membranes. Certain carbohydrates have globular shapes due to hydrogen bonding, which makes them extremely effective animal food stores. It gives the rigid, linear molecules of other carbohydrates the properties that make them ideal for use as structural elements in plants. Our DNA strands are held together by intermolecular forces, which also give our cell membranes structure, prevent water from boiling at room temperature and under normal pressure, and act as the physical adhesive forces that bind our cells, bones, and tissues.

The synthetic system being developed and natural bone formation both heavily rely on intermolecular interactions. Bones develop in living systems by specialized cells adhering to a long, fibrous natural substrate known as collagen. Along with collagen, some functional groups support the adhesion of bone-forming cells, while other functional groups support calcium crystallization. Tubules are formed when hydrophobic alkyl tails of molecules self-assemble into dispersion forces. The scientists added functional groups that increase cell adhesion and more functional groups that promote calcium crystallization to the other end of the molecule. They added functional groups that allow molecules to form covalent bonds with their neighbors after self-assembly, further stabilizing the initially non-covalent structure. It is useful to incorporate all of these properties into the molecular structure since the self-assembled fiber encourages calcium crystallization along its axis, much like the collagen found in nature. This molecular design is an interesting advancement at the nexus of nanotechnology and bioengineering. An easy, quick, and non-destructive experimental technique that can confirm the existence of different functional groups is infrared (IR) spectroscopy. One of the first steps you would take if you had a sample of unknown identification is to collect an IR spectrum, along with information on its solubility in common solvents, melting point, and/or BP. Like all types of spectroscopies, IR spectroscopy relies on molecules or atoms interacting with electromagnetic radiation. Organic compounds' atoms and groups of atoms vibrate more intensely around their covalent bonds when exposed to IR radiation. Since functional groups contain distinct configurations of bound atoms, an organic molecule will absorb IR radiation at frequencies that are unique to the particular functional groups that make up that molecule.

An IR beam is sent through a sample in an IR spectrometer, (Figure 1.21) which compares the radiation emitted via the sample to the radiation transmitted without the sample (Alvarez-Ordóñez & Prieto, 2012). The difference will reveal any frequencies that the sample absorbed. The results are plotted by the spectrometer as a graph of absorbance against frequency or wavelength. Two things can influence the frequency of a certain stretching vibration in an IR spectrum. Lighter atoms vibrate at higher frequencies than heavier ones. Double bonds are harder than single bonds and vibrate at higher frequencies than triple bonds, which are harder than double bonds. It is extremely unlikely that two different substances will have the same IR spectrum because even the IR spectra of relatively basic compounds have so many peaks. Because of this, an IR spectrum has been referred to as a fingerprint. As a result, with organic molecules, one can be positive that two pure samples with differing IR spectra are of separate substances. They are almost certainly the same substance if they exhibit the same IR spectra. The OH stretching absorptions of the hydroxyl groups of alcohols and phenols make it simple to identify them in IR spectra. These connections directly demonstrate hydrogen bonding.

The range of an alcohol or phenol OH group's IR absorption ranges from 3,200 to 3,550 cm^{-1}. The molecules' interaction by hydrogen bonding, which results in a wider distribution of stretching frequencies for the OH bond, is what causes the peak to typically be broad/wide. O-H absorption manifests as a steep peak in the 3,590–3,650 cm^{-1} range when alcohol or phenol is present in a solvent that cannot support hydrogen bonding. Due to the analyte's molecules being too far apart, intermolecular hydrogen bonds cannot form in very diluted solutions of this solvent or in the gas phase. Therefore, free hydroxyl groups are responsible for a sharp peak in the 3,590–3,650 cm^{-1} region. The sharp peak in the 3,200–3,550 cm^{-1} range is replaced by a broadband when the alcohol or phenol concentration rises. These effects are demonstrated by hydroxyl absorptions in the IR spectra of cyclohexyl carbinol (cyclohexyl methanol) run in both dilute and concentrated solutions. IR spectroscopy is a potent method for identifying functional groups.

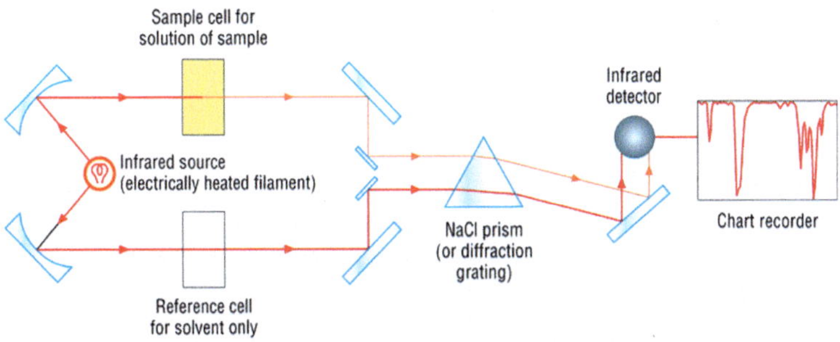

Figure 1.21. *Infrared spectrometer.*

Source: https://chem.libretexts.org/@api/deki/files/397516/IR-spectrometer.png?revision=1, Creative Commons License.

CHAPTER

2 Physiochemical Properties of Organic Compounds

Contents

2.1. INTRODUCTION

Understanding the physicochemical characteristics of organic molecules is essential for many of the tasks carried out by chemical engineers and scientists (Roman et al., 2019). Such a problem can include predicting a chemical's bioactivity, bioavailability, behavior during chemical separation and dispersion between environmental segments. Bioactive compounds like drugs, biocides, industrial waste products and chemicals, poisons in natural waters, and air pollution are causes for concern. Unfortunately, there is either no experimental data at all or very little of it for the great majority of the thousands of organic chemicals that are produced and routinely discharged into the environment. In the United States, the Toxic Substances Control Act (TSCA) inventory consists of about 60,000 items, and it grows by 3,000 each year (Zeeman & Gilford, 1993). The Environmental Protection Agency (EPA) of the US receives 3,000 compounds for the premanufacture notification process, the bulk of which lack experimental support (Creager, 2021). For the many compounds that lack experimental data, the only way to take accurate measurements is to estimate values using approximating/experimental methods. Estimated values may be reliable enough to classify substances based on vital characteristics. Researchers can qualitatively predict how substances would behave in environmental systems during waste treatment, chemical analysis, or bioavailability.

2.2. QUANTITATIVE PROPERTIES

Quantitative property-property relationships (Figure 2.1) are mathematical associations between the query property and one or more other properties (QPPRs) (Debia & Krishnan, 2010). QPPRs can theoretically be produced using physicochemical principles, as well as practically using experimental data and statistical techniques. On the other hand, quantitative structure-property relationships (QSPRs) connect a molecule's structure to numerical representations of physical properties. The molecular structure is a qualitative characteristic by nature, thus before correlations can be evaluated, structural information must be expressed into numerical values, known as molecular descriptors or indicators. Molecular descriptors are derived from the compound structure using structural information, fundamental or empirical physicochemical constants and correlations, and stereochemical principles. The molecular mass is an example of a molecular descriptor. It is derived from the molecular structure of the molecule and the atomic masses of its constituent atoms. A crucial chemical concept that aids in evaluating characteristics is similarity in structural. The fundamental principle of the theory is that a compound's properties are determined by its chemical structure and that molecules with similar chemical structures. Properties, which can be expressed qualitatively or quantitatively, can be used to describe compounds. Most physicochemical properties often depend on one another to variable degrees or having some form of interaction with one another. This cognitive approach is often slow and limited to a small number of chemicals when used by people as opposed to machines. Additionally, it lacks quantitative rigor. Using computerized algorithms, it has been possible to calculate the structural similarity of hundreds of molecules swiftly, identify structural discrepancies, and evaluate the relationships between structure and features. Several methods have been developed to translate molecular graphs into a computer readable language suitable for the evaluation of chemical structures, such as the identification of chemical structure similarities.

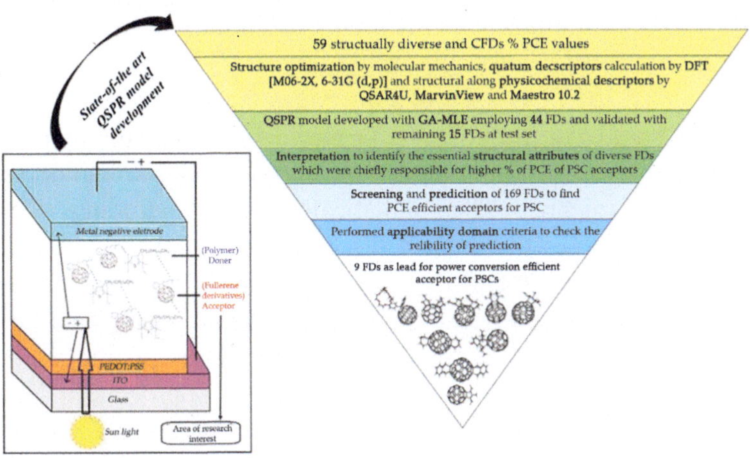

Figure 2.1. Quantitative property-property relationships.

Source: https://ars.els-cdn.com/content/image/1-s2.0-S2211285516301914-fx1.jpg, Creative Commons License.

Chemical property estimation is the process of extrapolating a query compound's unidentified property from known characteristics, molecular descriptors, or reference compounds. The query determines the subset of reference compounds that will be used as the training set. Training sets could include homologous and structural isomers as well as other types of closely related but strictly specified molecules. Toluene and benzene are used as examples, and chlorobenzene is used as the query compound in a bigger data set of n measured components. The training set is made up of the n compounds with known K_{ow} octanol-water partition coefficients. The property/property relationship that connects S_w to K_{ow} can be derived using the K_{ow} data set and the water solubility S_w data set. Benzene, toluene, and chlorobenzene are three molecules that are comparable to one another since they are all hydrophobic and have low molecular weights. The process of solubilization in water is comparable to partitioning in octanol-water in that the solute in both situations distributes itself between a polar phase (water) and a non polar phase. The link between K_{ow} and S_w is referred to as a quantitative property/property relationship since it connects two distinct properties.

To calculate the K_{ow} of the chemical chlorobenzene, the QPPR is employed. Similar to this, by analyzing the connection between molecular descriptors and a property, a training set can be utilized to create a structure/property relationship. A link between K_{ow} and the molecular descriptor X is established using a training set of K_{ow} data. Quantitative structure/property relationships are these relationships are known as (QSPR) (Fernandez, Woo, Wilmer, & Snurr, 2013). The K_{ow} of a query can therefore be estimated using this QSPR. Naturally, the training set must have a minimum number of entries and the properties of the compounds represented must span an appropriate property range in order to produce statistically significant findings. Rules that enable the prediction of the impact of structural alteration on a compound property have been developed for a small number of isomeric groups and homologous series. Typically, compounds and characteristics that lie within the range provided by the training set used to build the specific relationship are only eligible for usage with QSPR and QPPR approaches. Divide the structure into chemically logical components, such as groups of atoms, and assign each group a contribution to the property of the entire molecule. This is another often-used technique to discover empirical connections between structure and functionality. The group contribution model (GCM) is the name given to this method. To statistically assess each group's contributions, the properties of compounds containing each individual group as a component of a molecule are compared. The K_{ow} is produced as the result of the addition of the group contributions from the phenyl group and the chlorine atom. The foundation of similarity-based methods is the idea that closely related molecules will share similar characteristics. These methods start with one or more compounds (the k nearest neighbors, or kNN) with known properties that are closely related. The property value of the query is then further approximated using a model, such as averaging or a GCM. The kNN technique has traditionally been utilized in semi-quantitative or categorical property estimation. Toluene has been identified as a substance similar to chlorobenzene.

Often, it's crucial to understand both the property's temperature dependence and the property itself at a standard temperature. Many different fluid substances, including solvents, have temperature functions available. These activities, though, are compound-specific. Functions that describe characteristics as a function of both molecular structure and temperature have been created for a small number of substances. If 2 molecules have the same molecular formula, they are said to be isomeric. In some ways, all molecules with the same molecular formula belong to the same collection of structural isomers. However, 1-butanol and its five structural isomers could have different chemical compositions. When these molecules are arranged in a row, any two of them produce two isomers. It is useful to further categorize the isomers according to the kind, position, and branching of the functional groups for the purpose of property estimation. Positional isomers differ from one another by the location of a functional group inside a given molecule. The location of the hydroxyl group is denoted by the prefixes 1 and 2, respectively, making 1-butanol and 2-butanol positional isomers. As seen by their names, 2-oxapentane, and 3-oxapentane, where the prefixes 2 and 3 denote the position of the ether group, respectively, methyl n-propyl ether and diethyl ether are also positional isomers. The degree of branching in the alkyl groups of branched isomers varies. It's common to refer to the unbranched isomer as a regular isomer. The typical isomer only contains primary and secondary C atoms, which correspond to the methyl and methylene groups, in place of the atoms that make up the functional group. Branched isomers, on the other hand, have tertiary and/or quaternary carbon atoms.

2.3. ISOMERS

Isomer molar masses are equal by definition for an isomer. Significant correlations exist between numerous attributes and molar mass. Therefore, features of isomeric molecules within this class should be roughly comparable. Such generalizations should be used very sparingly, though. For instance, although isomers, anthracene, and phenanthrene have water solubilities that differ by a factor of roughly 100. As was done with tetrachlorobenzyltoluenes (TCBT), in some circumstances the qualities for a group of isomers are well described in terms of a property interval and a mean isomer value. A family of positional isomers known as TCBTs are molecules that cannot be superimposed by rotation or translocation. Diastereomers are stereoisomers that are not enantiomers. Examples of diastereoisomers are cis- and trans-1,2-difluoromethane. The spatial separations between the atoms that make up an enantiomer and the separations that make up its enantiomeric counterpart are pairwise identical. Because of this, two enantiomers will have similar molecular characteristics with the exception of how they interact with other stereoisomers and light. The development of separation techniques to identify enantiomers and produce samples of high purity is made possible by selective molecular recognition, such as that caused by a receptor or biocatalyst. There are significant biological and environmental ramifications to this particular stereoisomer interaction. Enantiomeric selectivity and purity affect a drug's toxicity and efficacy. For instance, the racemic combination of the sedative thalidomide that was supplied to expectant mothers caused birth abnormalities in the infants, whereas the pure R-enantiomer had no such problems. Diastereomeric pairs (Figure 2.2) do not all correspond to the same spatial distances, unlike enantiomeric pairs. For instance, the F-F and H-H lengths between cis- and trans-1,2-difluoromethane are different (Ruiz, Sears, Pines, & Jameson, 2006). As a result, diastereomeric compounds have variable energy contents and characteristics. The physicochcmical characteristics of cis- and trans-1-phenyl-1,3-butadiene serve as an example of the differences between diastereomers.

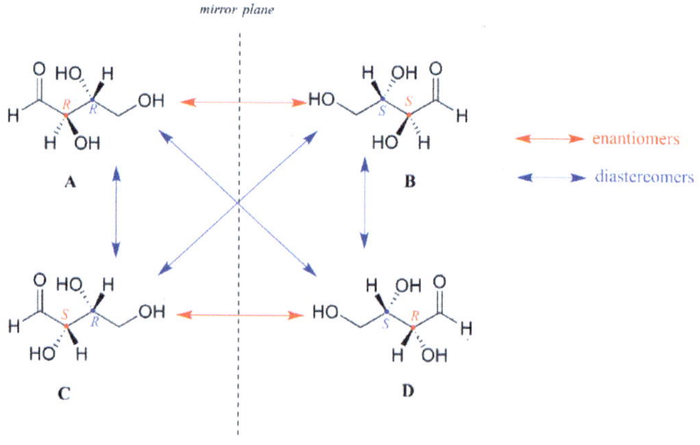

Figure 2.2. Diastereomeric pairs.

Source: https://chem.libretexts.org/@api/deki/files/141265/quatrose2.png?revision=1&size=bestfit&width=732&height=456, Creative Commons License.

For specific circumstances, it is known that a given molecular formula can have a very large number of isomers. The number of potential alkene and alkyne isomers has been examined by Coffman, Hense, and Blair. A group of homologous compounds is made up of successive components that differ exactly by multiples of CH_2 in their chemical structure. A homologous series is the name given to such a set. Similar to one another, homologous compounds have the same basic carbon skeleton with the exception of one or more added (or removed) methylene groups. Rules that anticipate the attributes of homologous compounds are based on compound similarity since the incremental contribution of a methylene group to a property is negligible compared to the contribution of the parent. Phase Shift Homologous relationships frequently apply exclusively to constituent substances with the same physical state. Relationships that apply to compounds in their liquid state are the most prevalent. The odd-even effect describes how some qualities depend on how many carbon atoms are present in a molecule. The melting points (MPs) against Ne are a common example of the odd-even. For a variety of series, including alkanoic acids and their anhydrides, alkyl alkanoates, alkyl p-nitrobenzoates, and mono and dialkyl ethers of stilbene, similar graphs have been reported. For features other than melting point temperature, such as transition point properties and solubility behavior, researchers provide instances of the odd-even phenomenon.

2.4. SOLUBILITY

The estimation result must also be qualified by pointing out any further potential model constraints. For instance, one must ensure that the query ingredient is a liquid if a model only applies to liquids. Water solubility is the sole attribute employed as the predictor variable. Because it has one aromatic ring, the subject, chlorobenzene, is compared to benzene and toluene. The methyl group of toluene is identical in size and hydrophobicity to the chlorine substituent. The applicability of the QPPR is extended to include all compounds that are similar to the set of n compounds that were included in the training set if the range of compounds is enlarged to additional compounds. QPPR models must be carefully chosen, taking into account the structure of the query compound and how it relates to the structures represented in the training set. Comparing the outcomes of using several models is frequently helpful. A molecular descriptor is calculated from the molecular structure, as opposed to a chemical property that can be measured. The atoms that make up the molecule and their spatial arrangement are included in the structural information. The geometric characteristics, such as the size and form of the molecule, can be inferred from the atoms' coordinates. Since the atomic masses are unaffected by the chemical bonds, they are a part of, the result is precise. However, the molecular mass (M) is a poor predictor for the majority of qualities since it only reflects a small portion of the geometrical and chemical characteristics of a compound. Connection tables that encode the chemical composition, including details about the sorts of atoms and bonds, make for better start points for building QSPRs. A collection of dependable guidelines can be used to obtain molecular descriptors from the molecular graph, or connection table, of the molecule. Common names for these descriptors include molecular connectivity indices (MCIs) (Lu, Tao, Hu, & Dawson, 2000). Other notations are also employed, including graph theory index, topological index, and molecular invariants. In terms of their statistical analysis and their applicability, these traits set QPPRs apart from QSPRs. It should be noted that specific characteristics of a query compound must be known in order for a QPPR model to estimate the property of interest. A collection of molecular descriptors is frequently applied in QSPR research in order to compare their importance for a given association.

The C_{sp} carbon's contribution is taken into account in both the single bond, =C-C=, and the triple bond, –C=N, when using the bond contribution approach. C_{sp} is a member of the functional group –C=N according to the function-oriented approach. Depending on the specific GCM scheme being utilized, the concept of functional groups may vary. Different GCM schemes have distinct subclassifications for groups with respect to their structural environment. Some GCMs make distinctions between equal groups based on how often they appear in a chain or ring. The substitution side of terminal groups, or substituents, such as an aromatic ring, nonaromatic ring, or alkyl chain, is frequently used to distinguish between them. There is no GCM standard, hence a rigorous examination of the aforementioned traits should be done before using a specific GCM. The starting point for property estimates in similarity-based and group interchange models (GIMs) is one or more similar compounds. There may be structural similarities and differences between the query and its parents. By substituting one or more of the query's fragments for a component of the parent, it is possible to construct the query from the parent. A replacement involves both an insertion and a deletion. Once the constants of the inserted and deleted fragments are known, the property estimate may be determined. The production of chlorobenzene from toluene is an example. In this transition, a methyl is removed, and a chlorine substituent is added. There can be a requirement for several insertions and removals. Group interchange (GI) is the term used to describe the process of creating a query. The GIM takes into account how various GIs may have an impact on the relevant property. A more complex example is the conversion of 4-bromo-5-phenylpentyl oxiran (II) into 6-bromo-7-phenyl-1-heptenene (I). The advantage of applying the GIM approach over the GCM approach is that the estimated value's accuracy is solely dependent on the experimental accuracy of the database compounds and the contribution uncertainty for groups, as opposed to the uncertainties of all groups that would be required to estimate a property with a GCM. The number of necessary group contributions and correction factors increases quickly, especially as a query compound structural complexity increases and values for a given correction factor might not yet be present in the GCM scheme under consideration. In these conditions, it is difficult to extract dependable property values using GCMs unless they are improved. The complexity imposed by the many components is greatly narrowed by the GIM technique. Only the factors that apply to the interchanged groups must be taken into account; any other elements specific to the MCS need not be taken into consideration.

2.5. KNN

The performance of the kNN methodology depends on the database contents and the specific similarity measuring method used. The ranking of database compounds in relation to any query molecule is made possible by measuring the molecular similarity, which quantifies the similarity between two compounds. The query and database substance molecular structures serve as the primary basis for this quantification and ranking process. If one thinks that the kNN compounds are the compounds for a query that is the most comparable in terms of both structure and property, then the mean value approach is justified (Cheng et al., 2007). But frequently, molecules that are structurally compared show significant variances in their physicochemical characteristics. As a result, adding GIMs will greatly enhance property estimation using the kNN technique. Researchers have shown how the kNN and GIM methods can be used together to estimate log K_{ow}. The kNN GIM approach needs to be further investigated in order to be used in future structure-property investigations and the automated creation of precise, integrated, and instructive estimation modules. At a reference temperature, the majority of QPPRs, QSPRs, or GCMs produce results. Estimating properties that rely on temperature is frequently limited to compounds for which the temperature coefficients in thermodynamic or empirical relationships have been determined specifically for the chemical. The invariants (constants) known as computed molecular descriptors, or X, are derived from the topological data. The type of atoms defined in relation to their positions and sometimes their relative types make up a molecule's topological information. These topological descriptors are related to particular chemical characteristics and activities.

The presence or absence of a specific structural or substructural property in the molecular graph is indicated by indicator variables, which are binary variables with a value of either O or 1. Some researchers incorporate indicator variables into QSPR equations that show correlations between property and descriptors, such as indicator variables. But indicator variables are unnecessary. A QSPR model can be divided into two independent models instead of utilizing an indicator, one for all the compounds for which the indicator variable is zero, and one for all the compounds for which the indicator variable is one. The number of bonds that must be cleaved in order to create an aliphatic structure is denoted by a molecular descriptor known as the cyclomatic number. According to graph theory, graph G is the least number of edges that must be taken out of G in order to convert it into G of an associated acyclic graph. The graph-theoretic equivalent of breaking a bond to open a ring is removing an appropriate edge. Charge indices and features like the hydrocarbons' dipole moment, alkanes and alkanols' boiling points (BPs), and the enthalpy of vaporization of alkanes have been correlated. Any physicochemical feature that can be computed from structure input alone can be a computable molecular descriptor.

As molecular size increases, gas viscosity decreases. For liquids, this trend is reversed since the viscosity rises with rising N_c in the homologous series. Similar correlations have been found between viscosity and molecular weight for alkanes, alkyl cyclopentane and cyclohexanes, 1-alkenes, alkylbenzenes,iso alkanols, isoalkanols, alkanones, alkanoic acids, and esters, encompassing a molar mass range between 30 and 300 g mol^{-1}. For a number of homologous series, correlations between viscosity and density and refractive index have been assessed; correlations between viscosity and BP and between viscosity and vapor pressure, for instance, have been reported for n-alkyl, B-ethoxypropionates. The Porter equation represents the relationships between viscosity and vapor pressure; the van Velzen technique only requires molecular structure input. Temperature coefficients are the additive parameters associated with terms for functional groups and different corrections for configurational factors. The temperature at which vapor pressure is equal to one atm is known as the normal BP (760 mmHg).

The vapor pressures of liquid and solid substances fall between 0 and 1 atm depending on the surrounding environment. For high-boiling substances with large molecular sizes and/or a high level of molecular self-association, pressures that are almost zero are reported. Solvatochromic relationships are multivariate correlations between a property, typically a property of solubility, and solvatochromic parameters, which are parameters that take into consideration the interaction of the solute with the solvent (Kamlet, Abboud, Abraham, & Taft, 2002). The solvatochromic parameters for vapor pressure only need to take into account intramolecular interactions such self-association between the molecules of the solute.

2.6. VAPOR PRESSURE

The UNIFAC group contribution approach was used to create an estimating method for pure-component vapor pressures (Asher, Pankow, Erdakos, & Seinfeld, 2002). The created model is applicable to hydrocarbons, alcohols, ketones, acids, and chloroalkanes with molecular masses of less than 500 and in the vapor pressure range of 10 to 2,000 mmHg. This model was used to include PCBs and other chlorinated aromatic substances. The temperature at which a vapor pressure equals the atmosphere's pressure on the liquid is known as the BP. The BP at 1 atm is known as the normal BP. The boiling temperature can be altered by impurities in the liquid. Because most compounds decompose at higher temperatures, reported experimental T_b values are often lower than 300°C. Compounds with high T_b are distilled at a lower pressure. The temperature at which a chemical changes from its solid to liquid state for a specific pressure is known as its melting point, or T_m. The solid phase and liquid phase coexist in balance at the melting point. Sometimes the term normal melting point is used to describe the melting point at 1 atm. However, when the melting point is mentioned, it typically refers to the usual melting point. The terms melting point and freezing point are commonly used interchangeably. The direction in which the two approaches equilibrium differs. Due to a chemical reaction that takes place when a molecule is exposed to an increase in temperature, the melting point of some compounds may not be quantifiable. It is fair to argue that more compounds have melting point data available than any other observable attribute of a chemical.

Even though the majority of experimental solubility data have been determined in distilled, salt-free water, natural water typically contains a variety of anionic and cationic species of mineral salts that alter the electrolytic property and, consequently, its ability to dissolve organic compounds. The presence of functional groups capable of hydrogen bonds, such as hydroxyl and amino groups, is crucial for aqueous solubility. The solubility behavior of isomers with the same functional groups is predicted to be identical. In order to create predictive solubility models, researchers compare the use of artificial neural networks with regression analysis methods. They claim that the neural network model outperforms the regression-based model in terms of performance. For the model to take into account the structural variability in the training compounds, a wide range of molecular descriptors is required. The water solubility of a molecule decreases when a methylene group is inserted into it, but not in a way that can be predicted by existing GCMs. The difference between low Nc members of homologous molecules is where this impact is noticeable. A solute water solubility is greatly influenced by the intramolecular group interaction. It would be preferable to have GCMs with a collection of highly distinct groups that mainly account for their structural group. The number of compounds that simultaneously include defined groups and have measured data available now set a limit on the creation of such GCMs. A compromise between a precise, statistically sound model and a less precise model with substantially larger applicability must therefore be sought in GCM development.

2.7. SALTS

Inorganic salts are dissolved in seawater. A typical model solution for seawater is an aqueous solution containing 35 gL^{-1} NaCl. Researchers have conducted thorough investigations into the salt effect on the solubility of nonelectrolyte organic molecules. S_w, the solubility in pure water, and the solubility at various salt concentrations are correlated with one another depending on the molecule. For instance, depending on the specific structure of the PAH, the solubility of PAHs in seawater, S_{sw}, is between 30% and 60% lower than their solubility in freshwater. Molecules of organic compounds exhibit size and shape features that are explained by graph-theoretical constants like MCIs and ICIs. Indicator variables or dividing the model into several equations, each of which applies to a class of compounds with a specific structure, are two ways to add functional properties like the ability to form hydrogen bonds into structure-property models. By modeling K_{ow} as a function of several graph-theoretical constants for classes of compounds, each of which contains compounds with an equal amount of hydrogen bonds. Models do not necessarily replace currently accessible techniques like K_{ow}

GCMs. Models have the advantage of being error-free for any arbitrary structure in terms of graph-theoretical constants. The relationships between K_{ow} and a few molecular characteristics, including chlorine number, MCIs, van der Waals volume and area, molecular volume, and polarizability, are important. These models enable estimation using molecular structure input for a variety of substances. By using the CNDO or complete neglect of differential overlap approach, it is possible to extract molar polarizabilities from molecular orbital (MO) calculations (Rhee, Metzger, & Wiygul, 1982). For polar compounds like alkanols, alkanones, dialkyl ethers, and alkane nitriles that either include hydrogen-bond-accepting or hydrogen-bond-donating groups, some correlation has been discovered. The effectiveness of an ANN to predict K_{ow} from descriptors generated from quantum chemistry has been researched. A total of 302 compounds were used for the model training. Around 21 chemicals that weren't in the training set were used to test the model (Asmare, Nitin, Yun, & Mahapatra, 2022).

The atom contribution approach accounts for conjugated double bonds. Depending on whether they are C atoms or heteroatoms, the environment differentiates between separate atom kinds. The C-atom environment is constrained to the nearby bonds and the H atoms that are linked. The environment also contains non-hydrogen neighbors for heteroatoms. Compound hydrophobicity is raised by intramolecular hydrogen bonding. Some researchers compared the measured log K_{ow} and the estimated K_{ow} of 500 compounds capable of generating intramolecular hydrogen bonds to assess the impact of intramolecular hydrogen bonding on compound lipophilicity. The atom contribution approach has been tested for contributions from heteroatoms like O, S, Se, N, P, F, Cl, Br, I as well as H and C atoms. The hybridization state, oxidation number, and heteroatom environment of the C atom that the considered H atom is attached to determine the classification of the H atom. According to their hybridization and whether their neighbors are heteroatoms or carbon atoms, carbon atoms are categorized. Halogen atoms and the hydroxyl, ether, amino, nitro, and carboxyl groups are the substituents taken into account, and the contributions for each have been estimated using multiple regression analysis.

2.8. COMPUTER-AIDED EVALUATION

The computer-aided structural evaluation (CASE) methodology (Figure 2.3) serves as the foundation for the Klopman and Wang method (Lemmen & Lengauer, 2000). This method uses only the groups that stepwise multiple regression analysis determines to be the most important groups contributing to K_{ow}, as opposed to using all groups in the estimate scheme. Depending on the pH of the surrounding environment, polar substances may ionize. The sorption of these chemicals is influenced by the level of dissociation. Weak carbonic acids that are nonionized adsorb more strongly than comparable anions. Stronger adsorptive forces are observed for weak bases in the protonated form compared to the unprotonated form. Basic chemicals including pyridine, quinoline, and acridine have higher single-solute solubilities in acidic soil, which is indicative of stronger protonated form solubilization. For a variety of classes of chemicals, connections between structure and soil water partitioning have been shown using MCIs. For computer-aided examination of molecular structures, the simplified molecular input line entry system (SMILES) is often employed (Sushko et al., 2011). SMILES is well-known and computationally effective because it uses atomic symbols and a simple set of guidelines.

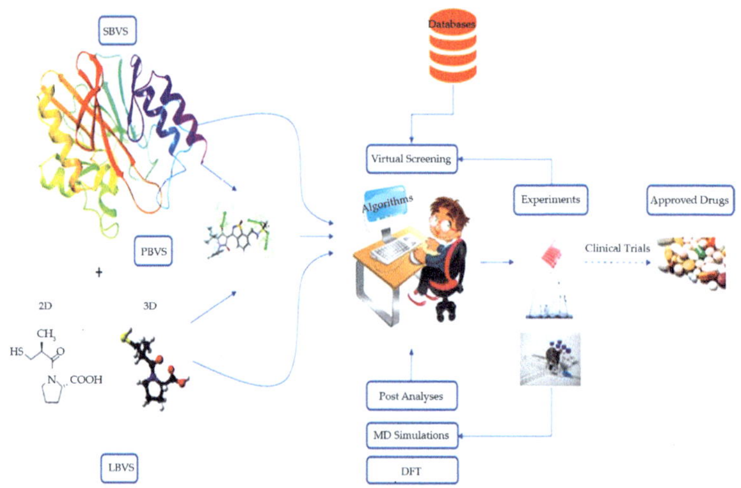

Figure 2.3. Computer-aided structural evaluation.

Source: https://ars.els-cdn.com/content/image/1-s2.0-S0223523421005547-ga1_lrg.jpg, Creative Commons License.

CHAPTER

3 Photochemistry

Contents

3.1. INTRODUCTION

All areas of research, including chemistry, biochemistry, medicine, biophysics, material science, analytical chemistry, signal transmission, and others, now include photochemistry (Figure 3.1) as a fundamental component (Feng & Nansheng, 2000). We are surrounded by products in our daily lives that use photochemistry or photophysics to execute their function or are manufactured with the help of these two fields.

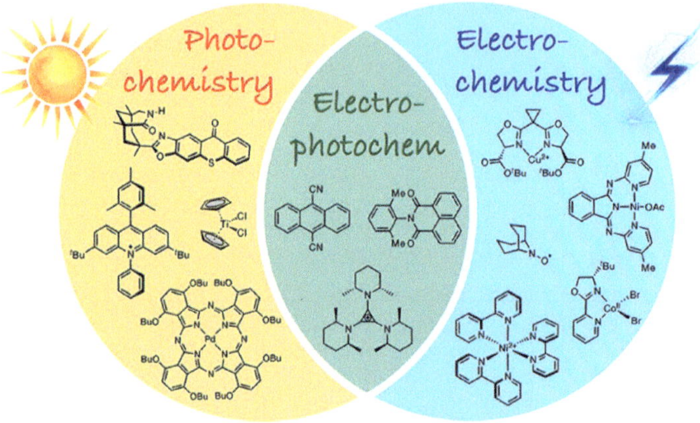

Figure 3.1. Photochemistry.

Source: https://pubs.acs.org/cms/10.1021/acscentsci.0c00549/asset/images/large/oc0c00549_0033.jpeg, Creative Commons License.

Examples include nanotechnology, sustainable technologies like solar energy, wastewater cleaning, security, and analytical devices like holograms, sensors, cosmetics for skin protection, hair coloring, etc., and information technology including computer chips and communication networks, data storage, displays, circuit boards, and LEDs. Fast spectroscopic techniques can be used to characterize the excited states and intermediates produced by pulsed excitation. It will be possible to identify single molecules, observe their reactions, and trace their motions in space using instrumentation with increasing sensitivity and spatial and temporal resolution. Electronically excited molecules have high reactivities and short lifetimes, which makes them extremely sensitive supramolecular structures because their reactions are likely to occur with reaction partners that are nearby when light is absorbed. Key photochemical reaction steps that may significantly reduce the number of reaction steps necessary to synthesize a given product are often avoided by synthetic chemists. With the growing focus on green chemistry, this can change. Electronically stimulated molecules have different chemical reactions from those in the ground state. The fundamental laws and principles of photoreactivity frequently go against those of ground-state chemistry.

3.2. SOLAR ENERGY

The majority of chemical changes in biota are driven by sunlight, either directly or indirectly. At a temperature of 5,800 K, the spectral distribution of sunlight (Figure 3.2) is quite similar to that of a black body (Kleidon, 2020). There aren't many alternatives in using sustainable technologies, particularly solar energy storage (photovoltaics, hydrogen production), nuclear fusion, wind energy, sea tide, and geothermal sources. Uranium ores that are worth exploiting are likewise few. In order to balance the global energy demand against our non-renewable resources over this century, significant efforts will be required to improve the efficiency of all processes needing energy and identify alternatives to energy-demanding behavior. Processes that are light-driven will play an important part in the transition to a low-C economy. Currently, the output of photovoltaic cells is growing at a rate of roughly 40% annually. When these methods can compete with the escalating expense of fossil fuels, significant increases in the use of renewable energy sources will follow. Using common laboratory procedures like quenching and sensitization, physical organic chemistry approaches have provided us the complete understanding of the characteristics and chemical behavior of transitory intermediates. The lifetimes of intermediates can only be estimated from data by making some assumptions about the rate constants of quenching or trapping. Working backward from the structure of stable photoproducts and the variation of their yield as a function of different additives, however, necessitates demanding preparative and analytical efforts. On the other hand, flash photolysis offers very limited data that enables the identification of the observed transients while producing absolute rate constants with little effort.

Since kinetic data produced using optical flash photolysis are susceptible to errors, combining some techniques is strongly advised. Light is first absorbed in the retina of the eye, which has 90 million rod cells and 5 million cone cells (Rieke & Baylor, 1998). There are three types of cones in humans, each with a unique response curve. Cones enable the perception of color even though they are less sensitive to light than rods, which facilitate vision in dim light. They react to stimuli more quickly than rods do. Maxwell made an attempt to chart the colors in a triangle made up of three main light sources.

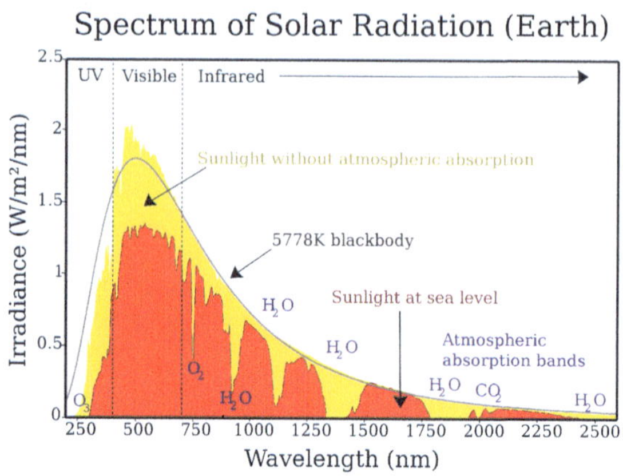

Figure 3.2. Sunlight.

Source: https://www.e-education.psu.edu/meteo300/sites/www.e-education.psu.edu.meteo300/files/images/lesson6/Solar_spectrum_en.svg.png, Creative Commons License.

Figure 3.3. Colors.

Source: https://upload.wikimedia.org/wikipedia/commons/e/eb/Subtractive-Additive-Colour-Mixing.jpg, Creative Commons License.

Red, green, and blue light (RGB) is typically used in the additive reproduction process to create other colors. The additive secondary colors cyan, magenta, and yellow are created by mixing one of these additive primary colors with another in equal proportions (CMYK, K for black) (Figure 3.3). The hue refers to the color that corresponds to this main wavelength. The term saturation refers to how much hue is there in the light. Saturation drops as white light dilutes the dominating wavelength or hue. When a dye absorbs at a certain wavelength, like 480 nm (blue light), the complementary color, yellow, is what is seen (white light minus blue). Subtractive colors are those created by paints, dyes, inks, and other materials. No dye that absorbs at a single wavelength can generate the color green as its counterpart pair since there is no spectral color purple.

Chlorophyll, (Figure 3.4) a green dye, contains two absorption bands, one about 450 nm and the other at 700 nm (Kay & Graetzel, 2002). The blue sky and the reddish hue of the sun at dusk are two illustrations of colors with low saturation. These hues result from the scattering of sunlight by Rayleigh in the atmosphere, which is caused by changes in air density and nanoscale particles. Tangential sunlight at dusk and dawn is reddish and lacking in short wavelengths (blue). The sky appears blue because sunlight that is refracted by the atmosphere is richer in short wavelengths. An extremely intricate physiological process underlies vision. The rods and cones, which are the main light sensors, are only one feature of the retina. Behind them, a sophisticated network of neurons processes the light signals before sending them to the brain.

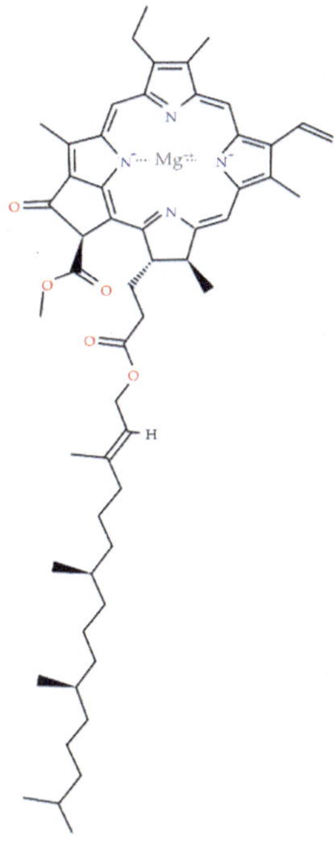

Figure 3.4. Chlorophyll.

Source: https://upload.wikimedia.org/wikipedia/commons/thumb/d/dc/Chlorophyll_a.svg/320px-Chlorophyll_a.svg.png, Creative Commons License.

3.3. FACTORS

Three factors produce magnetic fields in molecules: nuclear spin, electronic spin, and charge movement (motion) in space. To deal with photophysical processes that include a change in magnetic interactions, the only driving mechanism for such processes must be taken into account. The symmetry of electronic spin functions must always be considered explicitly due to the Pauli principle's influence on how electrons are distributed. The term vibrational relaxation refers to all procedures by which extra vibrational energy that a molecule has absorbed through a vibronic transition is discharged into the surrounding medium. It is also known as vibrational cooling or vibrational deactivation, or thermalization. It happens when molecules collide with one another. Thermalization of large molecules typically occurs with a half-life of a few picoseconds in solution but may take much longer in poorly conducting rare gas matrices at low temperatures. In solution, the time required to transfer a vibrational quantum to the surrounding medium is on the order of a vibrational period, which corresponds to the time elapsing between collisions with the solvent. In high vacuum, isolated molecules cannot relax their vibrations through molecular collisions, and infrared (IR) radiation is an inactive method of releasing excess vibrational energy. A heated molecule will nevertheless experience intramolecular vibrational redistribution (IVR), which distributes the energy that was once localized in the mode filled by a vibronic transition among the other vibrational modes (Kovalenko, Schanz, Hennig, & Ernsting, 2001). In this way, the molecule serves as its own heat source. In big molecules like benzene, IVR occurs on a time scale of < picosecond, but it may take much longer in small compounds with low vibrational state density.

3.4. PHOTOCHEMICAL REACTIONS

Primary photochemical reactions are ones that result in a primary photoproduct that is chemically distinct from the initial reactant and is started from an electronically excited state. The photophysical and photochemical processes that ultimately return the reactant to its ground state are continuously in competition with one another. Only by overtaking photophysical processes can photoreactions that result in new compounds be efficient. Consequently, it is crucial to have an understanding of the temporal frames of the latter. It is not easy to distinguish between photochemical and photophysical processes. The E-Z isomerization of alkenes is considered a photochemical process; but, in terminal alkenes, whose geometric isomers are indistinguishable in the absence of isotopic labels, the same process may be viewed as photophysical.

Fluorescence (Figure 3.5), phosphorescence (Figure 3.6), and chemiluminescence(Figure3.7)arealltypesofspontaneousradiationemission from electrically excited substances. The interaction of electromagnetic radiation with excited molecules results in stimulated emission, the microscopic counterpart of absorption, in addition to the spontaneous emission of excited molecules, fluorescence, and phosphorescence. Prior to the development of quantum mechanics, Albert Einstein deduced the existence of a close link between the rates of absorption and emission in 1917 (Parker, 2007). In radiation-less transitions, electronic energy from the initial, excited state is transformed into vibrational energy in the lower-energy electronic state. Radiation-free transitions between electronic states are not conceivable in the Born-Oppenheimer approximation, the potential energy surfaces act as solid barriers.

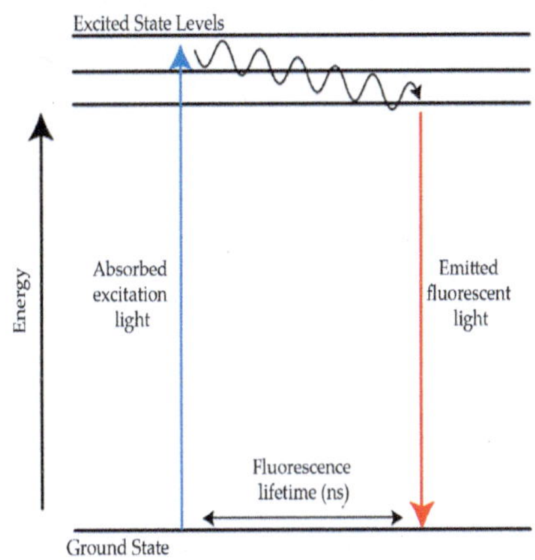

Figure 3.5. Fluorescence.

Source: https://upload.wikimedia.org/wikipedia/commons/thumb/5/53/Jablonski_Diagram_of_Fluorescence_Only-en.svg/800px-Jablonski_Diagram_of_Fluorescence_Only-en.svg.png, Creative Commons License.

Organic molecule photoreactions frequently take place in the triplet state. As a result, intersystem crossing occurs twice: once when the excited singlet state S_1 gives rise to the triplet state, and once when the major photoproduct, such as a triplet biradical, returns to the singlet manifold to create a stable product in the ground state. The reaction's overall rate is frequently determined by one of these slow phases. Long-lived excited triplet states can be manipulated in terms of product distribution since they can be easily quenched. A spin flip, which is necessary for ISC, can only be produced via magnetic interactions. Electron spin has a weak magnetic connection to other types of motion.

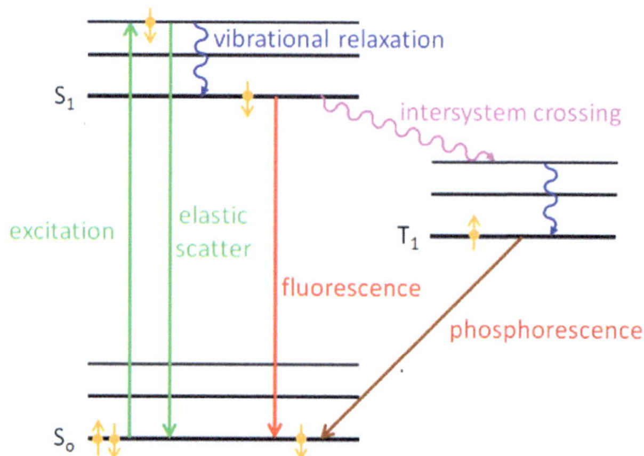

Figure 3.6. Phosphorescence.

Source: https://www.oceanopticsbook.info/packages/iws_l2h/conversion/files/Jablonski-fluor-phos-600.png, Creative Commons License.

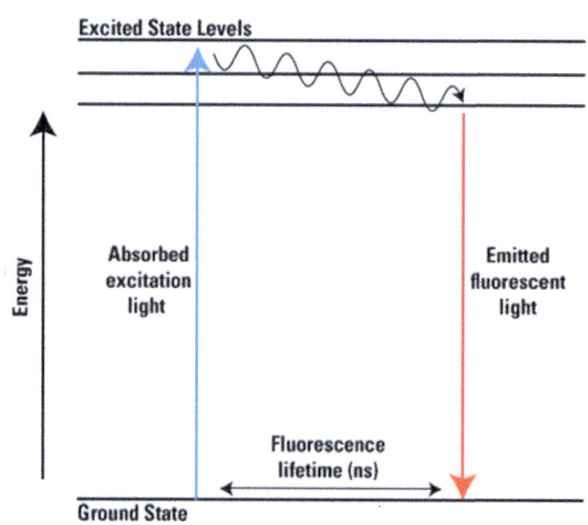

Figure 3.7. Chemiluminescence.

Source: https://www.sepmag.eu/hubfs/Chemiluminescence%20vs%20Fluorescence.png, Creative Commons License.

As a result, ISC rates are typically lower than IC rates. In order to capture the weak magnetic interactions, spin-orbit coupling (SOC) must be incorporated in the Hamiltonian operator. Hyperfine coupling, the interactions of the electron spin with the magnetic field of the nucleus, may become predominant in biradicals with spatially separated radical centers. According to the Franck-Condon principle, an electronic transition occurs without the nuclei's locations in the molecular entity or its surroundings shifting. Atoms move far more slowly than electrons, which are much lighter. Therefore, it is possible to think of the change in electronic structure brought on by electronic stimulation as happening in the Coulombic field of frozen nuclei. The transition is known as a vertical transition, and the resulting state is known as a Franck-Condon state. It is an electronically excited state that is born with the structure of the ground state. Fluorescent whitening agents (FWAs) are industrial products that are frequently used to cover up the yellow tinge of paper, clothing, and road markings. They are utilized in polymer sheets that focus solar light (Leaver & Milligan, 1984). FWAs are colors that fluoresce in the visible spectrum, typically in the blue, and absorb in the near ultraviolet. Fluorescent dyes are used as marker inks to highlight text or to improve the visibility of signposts on ski slopes since they absorb and emit light at longer wavelengths. The intense fluorescence of a fluorescein solution gives the appearance of green, yet when the light source is seen through the solution, it is actually yellow.

3.5. APPLICATIONS

As optical brighteners are often only used for security markings on banknotes, checking for fluorescence is a standard way to identify counterfeit money. In addition to the other characteristics needed for industrial application, such as good adsorption and strong adhesion to the substrates, low toxicity, and rapid environmental degradation, desirable properties of FWA include high absorbance in the near-UV region, but not in the visible region, that is, a strong absorption band rising sharply below 400 nm, a high fluorescence quantum yield and consequently low quantum yields of ISC and IC, and high photostability. Stilbenes and styrenes, triazines, benzoxazoles, coumarins, naphthalimides, and covalently linked or fused combinations thereof are the most often employed chemical classes as FWAs. One of the most used FWAs in laundry detergents is DSBP, a distyryl biphenyl derivative, with an estimated 3,000 tons of production annually. Studies have been done on the photochemical breakdown of it by Fe(III) ions in aqueous media.

Electronically excited states and several of their principal photoproducts have substantial reactivity. Diffusion-controlled reactions, also known as intermolecular processes, which take place whenever reactant molecules in solution interact, are therefore rather typical in photochemistry. Because the majority of the solvent molecules impede the translational motion, the frequency between solute molecules is lower than collision frequencies in the gas phase. In contrast, once reactants come into contact with one other in solution, they repeatedly collide because the solvent molecules in the area prevent them from being separated.

The reactive center receives the energy collected by the antenna, which may be made up of tens or even hundreds of chlorophyll molecules, and turns it into chemical energy that may be stored. Because internal conversion and vibrational relaxation from higher excited states will typically be far faster than any competing intermolecular energy transfer processes, we shall only take into consideration the donor's lowest excited singlet and triplet state. There may be exceptions in gases at extremely low pressure, in systems with slow thermal relaxation, or in systems with rapid energy transfer when local concentrations are high. The energy lost by the donor is recovered in the acceptor via isoenergetic energy transfer processes. An excited vibrational level will initially be occupied if the electronic excitation energy is lower. The excess energy is then quickly lost to the medium, making the energy transfer irreversible. A photon is spontaneously released by the excited donor and then spontaneously absorbed by the acceptor in the process of radiative energy transfer. Planck's law for the frequency distribution of a black-body radiator was derived with consideration for the transfer of radiant energy. Because it has a conceptual foundation and needs to be taken into account in all situations, it is sometimes referred to as trivial energy transfer. We shall demonstrate that, except in very large, diluted systems where intermolecular interactions are insignificant, it is typically not the primary energy transfer pathway.

3.6. SPECTROSCOPY

Recently, the free-energy surfaces that control protein folding have been successfully investigated using single-molecule spectroscopy (SMS) (Figure 3.8) (Tamarat, Maali, Lounis, & Orrit, 2000). Moving past ensemble measurements is a crucial advancement for FRET applications. As FRET is repeatedly turned on and off, a single, denatured protein molecule's conformational variation as it searches for its native structure is documented.

We would be able to capture the dynamics of individual polypeptide chains as they wriggle through the energetic environment, each molecule finding its unique path to the folded state. Similar to this, single-molecule FRET can spot enzyme activity by keeping track of conformational alterations brought on by substrate conversion. FRET-based SMS will continue to draw a lot of interest. When the excitation energy of the acceptor is a bit lower than that of the donor, assuring strong spectral overlap, the conditions of FRET from singlet donors to acceptors in the singlet ground state are ideal. Particularly at low temperatures, where the overlap is limited to the 0-0 transition, FRET between identical molecules is relatively moderate. When energy is repeatedly transferred between single crystals, it can do so over considerable distances until the excitation reaches an impurity trap/band with lower energy.

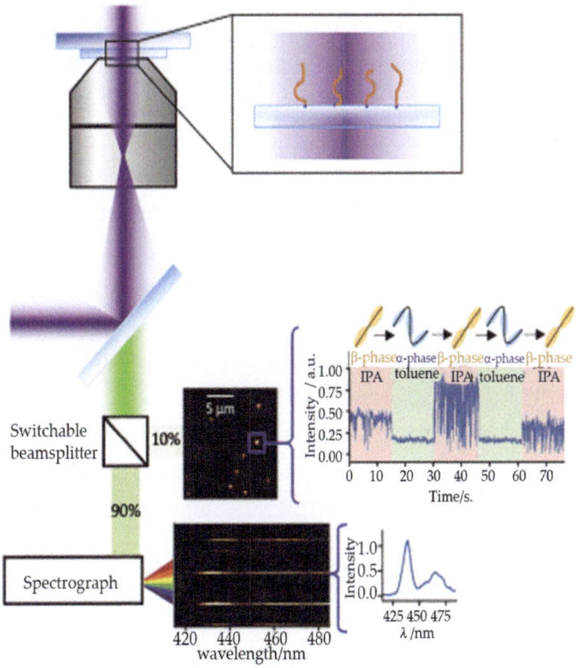

Figure 3.8. Single-molecule spectroscopy.

Source: https://ars.els-cdn.com/content/image/1-s2.0-S2590238519301249-fx1.jpg, Creative Commons License.

Interesting studies of FRET between chromophores in organized systems have been reported, including zeolites loaded with chromophores, monolayer assemblies, and self-assembling columnar mesophases, dendrimers, and other covalently linked chromophores that can be used as molecular switches, multiplexers, or antenna molecules. Simple models that solely take into account pairwise interactions between nearest neighbors may not be sufficient in such tightly coupled systems of m donor and n acceptor chromophores.

3.7. MECHANISMS

Studies on the temperature dependence of NVET rates in stilbene sensitization provided evidence in favor of a competing concept, the hot-band mechanism. According to this approach, access to geometries close to the relaxed excited state can be obtained by the thermal sampling of the ground-state acceptor's vibrational modes. It is corroborated by the observation that contrary to what would be predicted from Balzani's treatment, the energy differences across triplet donors are nearly exclusively represented in the activation entropy rather than the activation enthalpy. The triplet energy acceptor 1,3,5,7-cyclooctatetraene, which has a puckered structure in the ground state but relaxes to a planar structure in the triplet state, and a cyclobutadiene derivative, which transforms from a rectangular to a square structure, are two other examples of NVET. Continuing luminescence is displayed by lanthanide ions in the visible spectrum. However, because their f-f absorption bands are so weak, FRET to lanthanide ions is ineffective. By complexing with the right chromophores as an antenna, efficient FRET and even two-photon absorption can be accomplished. Lanthanide ion complexes are used for high-throughput screening assays to measure peptide dimerization associated with DNA transcription factors and ligand-receptor interactions as well as for imaging ion channels in living cells, studying protein-protein interactions or DNA-protein complexes in cells. Because the pyrene molecules are not linked in the ground state, time-resolved observations of pyrene fluorescence reveal that only the structured monomer emission is visible immediately after excitation. The broad emission then expands when the equilibrium between the monomers and excimers is reached, and the excimers are generated through diffusional contact.

First, we must keep in mind that entropy is always against the association of two molecules since one molecule loses its ability to rotate and translate. We can quickly determine the translational entropy of pyrene in the gas phase using statistical thermodynamics. Three fundamentally distinct forms of stabilizing interactions have been identified by quantum mechanics as being involved in the stabilization of excimers. Exciton interaction, which is the most noteworthy of these, can be thought of as the stabilizing energy brought on by the delocalization of the excitation energy between the nearby molecules. The fluorescence emission that has been discussed thus far is created by the direct excitation of a molecule to one of its excited singlet states. After IC to S_1, when an upper singlet state was initially populated, this excited singlet state emits fluorescence from S_1 with a span on the order of nanoseconds. Delay fluorescence is an emission that has the spectral properties of fluorescence but a span much longer than that of fluorescence, which can be produced by a number of methods that allow S_1 to regroup after ISC. These actions fall under the categories of P-type and E-type delayed fluorescence. G. N. Lewis proposed that biradical paramagnetic oxygen was a species in 1924. A molecular orbital (MO) model of dioxygen was presented four years later by R. S. Mulliken, who also predicted the occurrence of low-excited singlet states above the triplet ground state.

Kautsky asserted in 1931 that the photooxygenation of dyes was caused by the sensitive production of a transient, highly active, and diffusive form of molecular oxygen. He provided evidence for that assertion with a clever experiment in which a yellow dye and a colorless oxidation substrate (leuco-malachite) were separately adsorbed on silica gel beads and then combined. Only when the combination was exposed to oxygen did the colorless gel particles oxidize to produce Malachite Green; this process did not take place when either light or oxygen was present. For use in wind tunnel research, pressure-sensitive paint (PSP) allows for the viewing of two-dimensional pressure distributions on air-foil and model automobile surfaces (Gregory, Asai, Kameda, Liu, & Sullivan, 2008). A luminous pigment-containing polymer layer that is permeable to oxygen is applied on the surface of the air foil. Most commonly used are platinum porphyrins, which produce high phosphorescence that is severely reduced by oxygen. Using a CCD camera, the wind channel's surface is exposed to radiation. To make sure that no light can flow straight from the lamp to the detector, complementary filters are employed with the light source and the camera. Then, using a Stern-Volmer-type relationship for luminescence quenching, PSP provides a continuous map of the oxygen pressure on the surface as a function of wind speed. Porphyrin-sensitized singlet oxygen emission at 1,270 nm, which can be captured by an InGaAs near-IR camera, is the basis of a supplementary technique. The traditional approach to designing and optimizing the forms of air-foils and other surfaces is creating a load model by making hundreds of tiny holes called pressure taps, each of which is connected to multiplexed pressure transducers via a separate tube. The lift map required to properly engineer the support structure and other characteristics of an airplane is then provided by these pressure taps. These models take a long time to construct and can cost millions of dollars. With only a few pressures taps to calibrate the PSP emission, PSP aims to obtain the same information utilizing highly detailed digital photographs of luminescence. PSP can offer data from locations where taps cannot be installed, for as at extremely thin sections of a wing model.

The Franck-Condon principle states that a photon may initially be positioned on an elevated point of the excited state surface after being absorbed by a molecule. Because their location and shape can be determined from the vibronic structure of the absorption spectra, nearby minima on the excited state PES are known as spectroscopic minima. In a manner similar to how the potential energy of a rolling stone is finally transformed into heat by friction, the energy produced as the molecule begins to roll down on the excited-state surface is irrevocably dispersed to the surrounding medium. The molecule will probably follow the path of steepest drop like a marble. After reaching a local minimum, it is given the option of engaging in fluorescence, internal conversion, or intersystem crossover among other photophysical processes. Although this option is always available in theory, thermal relaxation happens most quickly on the PES's slope. Given enough time close to the minimum, heat reactivation may push the molecule over a barrier to find a deeper minimum on the excited PES. The latter's shape will determine what happens to it after it touches the ground-state surface. To forecast or estimate the morphologies of the pertinent PES as a framework for comprehending the photochemical reactivity of organic compounds, theoretical calculations, and models can be used.

Whether the excited-state surface represents a singlet, or a triplet state is not specified at this time. Internal conversion creates a vibrationally excited or hot molecule in the electronic ground state from the local/spectroscopic minimum in the excited state. For some reactions, the extra energy that IC or ISC initially gained is frequently more than enough. However, vibrational relaxation will quickly cool the molecule, and hot ground-state reactions that produce the new product infrequently outperform vibrational relaxation that returns to the starting point. In order to direct a significant portion of the extra energy into the internal degree of freedom that can heave the molecule towards during reaction, IVR is necessary. In other words, a barrier on the ground-state surface is sensed even at excess energies considerably above its height because the region above the barrier has a low state density. The fact that it is challenging to gather clear evidence for such a reaction path may be another factor contributing to the low number of hot ground-state reactions that have been confirmed to occur. On the other hand, vibrational cooling is moderate, and hot ground state reactions are extremely likely to happen in gaseous molecules at low pressure and in rare gas matrices at low temperature. Adiabatic reactions are those that take place totally on an excited-state surface and first produce the final product in its excited state. In the triplet state, adiabatic reactions are common because the competing photophysical decay processes move slowly, allowing passage through significant barriers on the triplet surface. Singlet state adiabatic processes, such as valence isomerization, geometric isomerization, and hydrogen or proton transfer only result in small structural changes with very low barriers. It was argued that Born-Oppenheimer states with identical spatial wavefunctions that share the same transformation properties under molecule symmetry operations cannot have exactly the same energy because an exact Hamiltonian operator would typically give rise to interaction terms that separate them. The non-crossing rule is only strictly true for diatomic molecules; surface crossings of polyatomic molecules can happen in a constrained, 3D geometric subspace. However, using a single response coordinate with only 1° freedom prevents inadvertent surface crossings at most geometries. Diabatic surfaces are approximate PESs that cross because a particular model utilized ignores the relevant interaction factors. It is mostly a matter of academic debate as to whether a molecule in a diabatic reaction actually descends through a conical intersection or a deterred crossing. When two states have very similar energies, PESs are useless because the Born-Oppenheimer approximation fails, and energy dissipation is extremely rapid.

3.8. PHOTOLYSIS

In the lab, photolysis is performed using synthetic UV or visible light sources in conjunction with optical filters that limit multiwavelength radiation to the desired spectral range (Wang, Zhang, Lou, Yang, & Chen, 2004). The mercury vapor pressure has an impact on their spectrum irradiance. The vapor pressure of mercury is around 103 mbar. Due to the de-excitation of Hg_3P_1 and Hg_1P_1, the former produces two bands of radiation with centers at 253.7 and 184.9 nm, respectively. The transmittance of the quartz envelope will determine how much of the short-wavelength line is filtered out. They are frequently used in Rayonet reactors. As a result of frequent collisions with electrons, the excited mercury atoms in medium-pressure arc lamps are partially excited to higher states. More emission lines that are helpful for photochemical investigations, such as 313.9 or 365.4 nm, become available as a result. The temperature of the lamp envelope can reach 600°C, and it generates a lot of heat and IR radiation. In order to prevent the sample from heating up, cooling-water circulation or water cuvette filters must be used. The high-pressure (100 bar) mercury arcs are the UV radiation source. Due to the high pressure and temperature, the spectral lines are widened and superimposed on a constant radiation background. The smoother xenon lamp spectrum finds value in environmental photochemistry research due to its resemblance to solar radiation, even if conventional mercury-xenon [Hg (Xe)] lamps still emit large mercury emission bands, especially in the UV area.

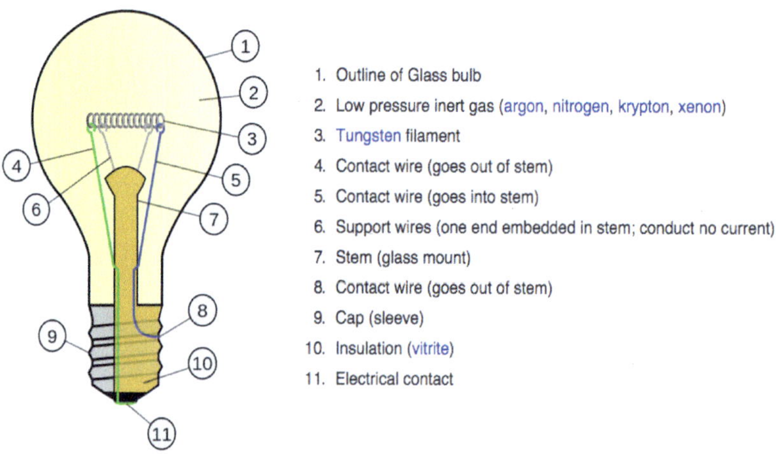

Figure 3.9. Tungsten-filament incandescent bulb.

Source: http://large.stanford.edu/courses/2014/ph240/dikeou2/images/22big.png, Creative Commons License.

Tungsten-filament incandescent bulbs (Figure 3.9) have a working temperature range of 2,200–3,000 K. They emit light in the visible and IR regions of the spectrum as a result. In photoreactions of colored chromophores, such as the photodissociation of bromine or chlorine molecules to start photo halogenation processes, radiation may be helpful. In photochemistry, lasers are primarily used for time-resolved measurements, but they can be very helpful for preliminary research or for illuminating samples in far-off locations like NMR or EPR cavities because they produce intense, monochromatic, parallel beams. Pulsed lasers can cause a single molecule to absorb two or more photons or re-excite intermediates, resulting in the formation of products that are distinct from those produced by continuous irradiation. When the photoproduct is also radiation sensitive and secondary photolysis of the product cannot be prevented by selecting a suitable wavelength, pulsed lasers may also be advantageous. Secondary photolysis can be totally avoided by irradiating each sample with just one laser pulse, provided that the photoproduct is not generated during the period of the laser pulse, a requirement that is frequently met when the reaction goes through reactive intermediates or a triplet state. The rotational speed is modified to match the laser's repetition frequency.

Low divergence monochromatic beams of coherent electromagnetic radiation are produced by lasers (Dutta & Manna, 2003). In 1960, Theodore Maiman created the first functional laser. Optical storage systems including CD and DVD players, bar code readers, laser pointers, and laser printers are currently used where lasers are used frequently. In photochemistry, lasers are used for cutting and carving, target detection, weapon delivery, surgery, and esthetic procedures. Due to their defined wavelength and brief pulse durations, they have emerged as the most important instrument in photophysical and spectroscopic research. An external energy source, often light from a flashlamp, or an electric discharge, is used to pump the medium. Only when there is an inversion, wherein more particles are in an excited state than in a lower energy state, does laser emission occur. This causes light to be amplified inside the cavity. A steady pump source keeps the inversion maintained in the continuous wave (CW) mode of operation, and the laser output is basically constant over time. Short pulses make up the output of the pulsed mode, and the cavity's length determines length/period between sequential pulses. By keeping the cavity output mirrors reflecting, the inversion is permitted to develop in a Q-switched laser. The reflectance is changed by an electro or acousto-optic Q-switch to release the light until the inversion is exhausted once the pump energy stored in the cavity reaches the required level. This results in pulse lengths that range from 3 to 20 ns. Mode-locked lasers produce pulses with an average duration of 30 ps that can be compressed to only a few femtoseconds utilizing various techniques. The energy-time uncertainty principle dictates that a pulse with such a brief duration must have a wide range of wavelengths, and the laser medium must have a broad gain profile to amplify them all. Artificially produced sapphire that has been titanium-dope is a suitable material (Ti:sapphire).

In a material with two electronic states, or a two-level system, inversion is not possible. When the ground state's absorption is countered by the excited state's stimulated emission, optical transparency results from a roughly equal proportion of the two states. It is necessary to populate the emitting excited state indirectly. In a three-level laser, the laser medium is irradiated to pump an upper level 2. This level is quickly drained by a nonradiative process. As a result, stimulated emission brought on by more pumping is avoided, and level 3's finally surpasses that of the ground state. For instance, benzophenone would meet these criteria, but it has a very low radiative rate constant of phosphorescence. Four-level systems can be used to build more effective lasers. Here, a low-lying level 4 is reached by the stimulated emission from level 3, which is swiftly depopulated to ground level 1 by a second nonradiative process. The majority of practical lasers are of the four-level variety because they are more effective than three-level lasers since just a small number of atoms must be stimulated into the upper laser level to create a population inversion between levels 3 and 4. Classical optics, often known as linear optics, makes the assumption that transparent matter will respond linearly to electromagnetic radiation as measured by its refractive index. Hooke's law, which describes the elasticity of a spring, is a well-known example. One GW, or the power of a contemporary nuclear power station, can be generated by a single 1 mJ pulse that is emitted within a picosecond. Due to nonlinear reactions and the high radiative powers of lasers, higher terms (such as cubic and quadratic) must be included. A BBO crystal's frequency doubling (second harmonic generation), optical parametric oscillators, and similar phenomena are examples of unusual but very useful nonlinear optics phenomena. However, nonlinear optics can also produce undesirable phenomena like dielectric failure.

3.9. QUANTUM DOTS

Quantum dot light (Figure 3.10) amplification research is now underway and has the potential to lead to innovative new technological advancements (Akiyama, Sugawara, & Arakawa, 2007). Heisenberg's uncertainty principle has the effect of making ultrashort laser pulses less monochromatic. Ti: sapphire femtosecond lasers have enabled the creation of optical frequency combs because they can generate an octave-spanning spectrum directly from the oscillator. An optical cavity's brief pulse is kept flowing by a mode-locked femtosecond laser. As a result of an attenuated copy of the light pulse escaping with each round, the laser consistently emits a train of ultrashort pulses. The beam and the pulse train are superimposed with a beam splitter to determine the laser wave's unidentified frequency, and a detector records the interference signal. Time and frequency metrology is extended by optical frequency combs from mode-locked femtosecond lasers. Sensitive limitations for potential slow fluctuations of fundamental constants are being established through precise comparisons of the optical resonance frequencies of atomic hydrogen and other atoms with the microwave (MW) frequency of a cesium atomic clock. Frequency comb technologies are being extended into the extreme ultraviolet area by optical high harmonic production.

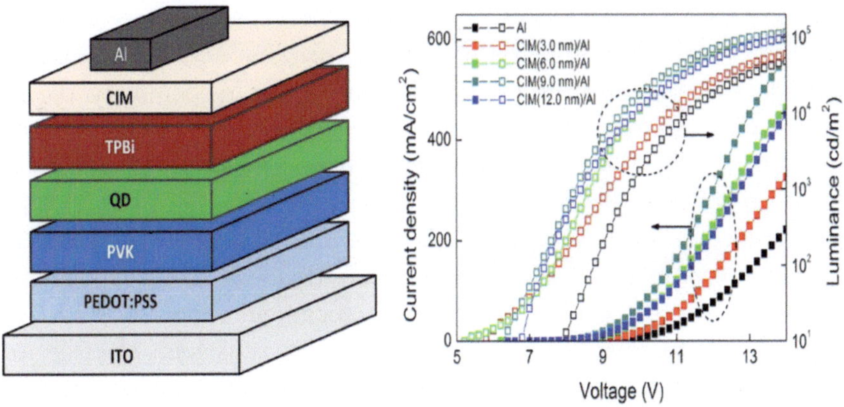

Figure 3.10. Quantum dot light.

Source: https://ars.els-cdn.com/content/image/1-s2.0-S1566119916300647-fx1_lrg.jpg, Creative Commons License.

Light sources having multiple sharp but widely spaced emission lines, like low-pressure mercury arcs, can produce basically monochromatic light and have their spectral output modified by optical filters. Cheaper alternatives to monochromators include easily obtainable glass filters and filter solutions. The glass components of the photochemical equipment, such as lamp envelopes, immersion wells, test tubes, and so forth, are the most typical cut-off (long-wavelength pass) filters. Between the radiation source and the sample, a glass filter can be added. Fused quartz and fused silica both have extremely high UV transmission; they cut off below 180 nm and, depending on thickness and quality, also partially below 250 nm, while becoming virtually completely transparent above 250 nm. While Pyrex is frequently used when wavelengths >290 nm are required, Vycor, a glass with strong temperature and thermal shock tolerance, can also be used for irradiation near and above 250 nm. Commercially available short or long-wavelength pass and bandpass filters come in a wide variety. Thin-film dielectric mirrors called dichroic filters selectively permit light with a limited range of wavelengths to pass. Dichroic filters don't absorb much energy; therefore, they don't heat up or deteriorate significantly over time when used for extended periods of time because undesired wavelengths are reflected rather than absorbed.

Any photochemical lab may easily prepare solution filters. Literature on their compositions and optical characteristics is widely available. Some of the described solutions, meanwhile, have limited photochemical stability and need to have their transmittance tested frequently. Pure solvents can also function as optical filters; for instance, benzene and methanol largely block light with wavelengths below 205 and 280 nm, respectively. Planning photochemical studies using solutions must take into account their (internal) filter impact because it may reduce the reaction efficiency. Single photons can be detected by photomultipliers because they are sensitive to light in the ultraviolet, visible, and near- IR spectrums. They are made up of a photocathode, a series of electrodes known as dynodes, and an anode inside of a glass or quartz vacuum tube. The photocathode material is struck by an incident photon with sufficient energy, which causes an electron to be ejected and directed toward the electron multiplier. This device is made up of about eight dynodes, each of which is maintained at a higher voltage than the one before it. More electrons are released upon impacting the first dynode, and these electrons are then propelled in the direction of the second dynode. At each stage, the dynode chain generates more and more electrons. The buildup of charge causes a brief current pulse that signals the arrival of a photon at the photocathode. Each incident photon can generate up to 108 electrons in a photomultiplier. The time resolution of photomultipliers is limited by the single electron pulse's temporal half-width, which is on the order of 1 ns when it reaches the anode.

In photodiodes, light absorption drives an electron from a semiconductor's valence band to its conduction band, resulting in the creation of an electron-hole and a mobile electron. A voltage will cause a photocurrent to flow through the diode when it is applied. A diode array is a one or two-dimensional matrix made up of thousands of photosensitive diodes. There is a maximum diode span of 100 mm^2. The charge needed to recharge a capacitor behind each irradiation diode [charge coupled device (CCD)], which is typically clocked at 30 Hz, can be used to determine the readout to store a digital image. In most photographic devices, light-sensitive films have been replaced with diode arrays. Microchannel plate (MCP) detectors, also known as image intensifiers, are a two-dimensional array of tubes that also work via the photoelectric effect (Tremsin & Vallerga, 2020). The MCP receives a projected image. A metal emits an electron that is accelerated in an electric field and amplified as a photon strikes a photocathode on a detector plate. When a single photoelectron strikes the anode, which is a phosphorescent screen, up to 10^4 photons are released. Because the human eye is most sensitive to green, green phosphor is typically employed. Thus, a monochromatic, amplified image is reproduced on the phosphor screen. MCPs are utilized in night vision equipment and have near-IR sensitivity. To record the image, the phosphor is frequently used in conjunction with a CCD camera. By adding a counter voltage close to the photocathode, MCPs can be gated for time-resolved applications with durations of roughly 1 ns. This prevents photoelectrons from entering the accelerating field, and the device remains opaque as long as the counter voltage is present. This can be used for spectrographic flash photolysis or to catch objects moving quickly.

The green phosphor displays red light on one side and blue light on the other due to the one-dimensional separation of the spectrum. An image can be read by a CCD camera using this method. Not the read-out of the CCD camera, but the gating of the MCP limits the time resolution. The reaction time of MCP photomultipliers is 30 ps. A streak camera converts a light pulse's temporal profile into a spatial profile on the detector. For instance, a one-dimensional photocathode array is struck by a spectrally defined light pulse from a fluorescent sample generated by a brief pulse of light traveling through a tiny slit in one direction. The electrons are deflected orthogonally to the slit after being accelerated in a cathode ray tube and passing through two plates that create a rapidly rising electric field. The spectrum distribution of the light pulse in one direction and the time-varying deflection of the electrons in the other direction are then displayed on the phosphor screen at the tube's end. The streak pattern on the screen is captured by a CCD array, exhibiting both the temporal and spectral characteristics of the light pulse. The streak camera needs to be triggered, just like an oscilloscope, to capture periodic occurrences. The fastest optoelectronic streak cameras have a time resolution of roughly 0.2 ps in single-pulse mode. The time precision of cumulative periodic signals will, however, be limited to 10 ps or more by the jitter of the source trigger and of the instrument response. LEDs are a constant presence in our daily life. On electrical gadgets, they are frequently employed as tiny indicator lights, and area lighting and other high-power applications are using them more. The color of the emitted light might range from IR to near ultraviolet depending on the composition of the semiconducting material utilized. Red LEDs were widely employed in seven-segment displays after they were made commercially available in the late 1960s. Later, a wide range of different colors became accessible and were used in numerous appliances. The light output of LEDs rose as the materials technology advanced, making them suitable to be used as lighting. Early in the 1990s, low-cost, efficient blue LEDs became available, completing the necessary RGB color trio. In the 1990s, high-brightness colors increasingly became available, allowing for innovative outdoor signage designs and massive video displays for billboards and stadiums.

3.10. DIODES

Organic light-emitting diodes (OLEDs) (Figure 3.11) have recently begun to be used commercially in bright displays, and the future is promising given their low price and performance (Dodabalapur, 1997). They might someday take the place of fluorescent and incandescent light bulbs as well as the backlit liquid crystal displays (LCDs) that are found in portable electronics like laptops and cell phones. The development of near-UV emitting LEDs and OLEDs has led to an increase in their use as light sources in photochemical apparatus for preparative irradiation and for quantitative work like quantum yield measurements. While complex apparatus and tools are required to research the processes of photochemical reactions, preparative irradiation procedures are comparatively inexpensive. However, several special conditions must be met when photochemical reactions in solution are exploited for organic synthesis. High chemical yields and a decrease in the formation of side products are possible, but photochemists must be careful when selecting the right irradiation wavelengths and reactant concentrations, take internal filter effects into account, and decide whether oxygen removal is necessary during irradiation. Since light is the chemical reagent (initiator), the reaction vessel and any optical parts (lenses) should be transparent to the desired wavelength range. The products or intermediates created during the process may absorb the irradiation's wavelengths. During the conversion, as their concentrations rise, the effectiveness of the reaction may drop and eventually cease. In some circumstances, side reactions can be avoided by lowering the reactant concentration, which also enhances light penetration into the sample, although this will necessitate using greater quantities of the reaction mixture. Dioxygen should often be removed from the solution before irradiation since it effectively quenches triplets. The most popular techniques involve either the more rigorous yet demanding freeze-pump-thaw process or purging the solution with argon or nitrogen for several minutes before or during irradiation. To completely remove all dissolved gasses, the latter method involves many cycles of freezing the solution in liquid nitrogen, vacuum-pump evacuation, and subsequent thawing when the vacuum input is closed.

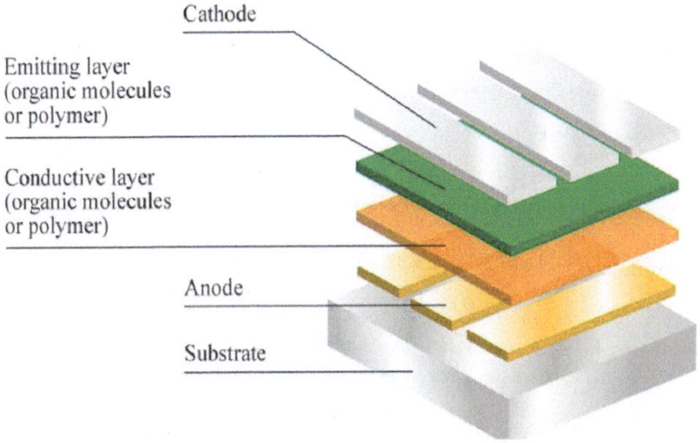

Figure 3.11. Organic light-emitting diodes.

Source: https://d3i71xaburhd42.cloudfront.net/a3b666d54e45f28ddc391adf95649da7ea383d36/2-Figure1-1.png, Creative Commons License.

The typical photochemical devices used in preparative irradiation are photochemical reactors. They typically consist of double-walled immersion wells with water cooling for the bulb and IR radiation filtration. The sample solution is positioned outside the lamp and is commonly agitated by a magnetic bar in a setting of argon. The reactors are not the best choice for photolysis experiments since those experiments need a narrow excitation band. The range of wavelengths that can be irradiated is constrained by the building material (Pyrex, quartz), although a filtering glass sleeve can be added between the lamp and the well wall. Polymeric compounds may precipitate on the reactor walls despite spinning and partially blocking light. An aperture allows for the continual removal of the photolyzed solution. The thin layers of reactors in photochemical processes allow for wide penetrability and, as a result, a more effective use of light. The solution is kept away from the radiation source in this setup. An external reactor consisting of the reaction sample put in the middle of an array of UV lamps is commercially available and can be used to regulate the temperature of the reaction mixture by submerging the entire setup in a water or methanol bath.

Electrodeless discharge lamps (EDLs) are used in a photochemical reactor that, when positioned in a MW field, produces UV radiation. The EDL comprises a glass tube filled with an excitable chemical and an inert gas and sealed under a lower noble gas pressure. Gas discharge is sparked by a low-frequency electromagnetic field (300–3,000 MHz), which results in the emission of UV-Vis light. The simultaneous UV and MW irradiation of the sample is one benefit of a setup in which an EDL is positioned inside the reaction solution, making it possible to perform photochemistry at high temperatures and at a cheap cost. Insoluble semiconductor metal oxides or sulfides are used in heterogeneous transition metal photocatalysis where they undergo electron transfer between the stimulated semiconductor surface and the reactants. The photolysis reactor typically consists of a transparent vessel with a reaction mixture (TiO_2 suspended in the reactant solution) that is vigorously stirred continuously. Similar to the photoreactor with an external arrangement, the source of visible or UV light is positioned above or adjacent to the reactor. Such equipment is used to purify wastewater on a massive scale. There are many different kinds of microreactors, including zeolites, silica, and alumina surfaces, macrocyclic host cavities made of ethers or cyclodextrins, and molecular aggregates of micelles or monolayers. The setups for photocatalysis experiments can be comparable to those for photolysis experiments. In the absence of any solvent, a solid that has absorbed reactants can also be exposed to external radiation.

When organic molecules are crystallized, many of them go through photochemical changes. Only specific conformations are accessible for the reactions because intermolecular interactions limit the rotational and translational motions of the molecules in the solid state. The crystalline substance is converted into a fine powder by means of an agate mortar and pestle and then put in a thin layer between two quartz plates in a photochemical experiment. The plates are enclosed inside a polyethylene bag that has been nitrogen/argon-purged before being exposed to radiation from an outside source. The emission spectra of the sun under various conditions, such as those at sea level or in space, are mimicked by a number of commercial suppliers through continuous and pulsed solar simulators. Optical filters are used to modify the wavelength output. In scientific and process control applications like photobiology, material stability testing, photolithography, solar cell testing, and environmental investigations, the simulators are frequently employed. Curing is the process of making a polymer material stronger or harder by cross-linking it with UV irradiation. This technology is a green method of processing inks and coatings that is based on an external UV irradiation of a solid surface. We can understand the displacement of the excited state potential energy surface with respect to the ground state from the vibronic structure of absorption bands. The total intensity of an absorption band, measured by oscillator strength, solvent shifts and substituent effects on the transition energy, and the orientation of the transition moment, determined by spectroscopy with polarized light, can be used to directly compare the predictions of quantum mechanical models for electronic excitation. The electronic structure, dipole moment, and reactivity of these excited states can be determined using quantum mechanical models.

Typical absorption spectrophotometers are double-beam devices that regularly compare the intensity of light passing through the sample to that of a reference beam. Due to the wavelength dependence of the monochromator transmission and detector sensitivity, this continuously delivers accurate yields for temporal fluctuations of the monitoring light source intensity as well as for its variation with wavelength. The intensity ratio of the probe and reference beams can only be accurately measured to a maximum of 0.01%, despite the fact that photomultipliers are extremely sensitive and are getting close to the limit of single photon detection. The majority of spectrometers display absorption spectra as a function of wavelength. Spectroscopists have traditionally chosen measures like frequency or wavenumber, which are proportional to energy rather than wavelength. The vibronic characteristics are evenly spaced, and the first absorption band frequently displays a mirror-image relationship with the fluorescence spectrum. Single-beam instruments are used in fluorescence spectrophotometers. Since emission spectroscopy is essentially a zero-background technique and photomultipliers can reliably detect single photons under ideal operating conditions, it is very sensitive. A single incident photon causes a photoelectron at the photocathode to emit an avalanche of electrons. The reported temporal resolution of photomultipliers does not apply to the period of time between the ejection of a photoelectron at the photocathode and the arrival of the pulse at the anode, but rather to the width of a single pulse. Electron heat release can be decreased by cooling the multiplier housing. Hot electrons typically produce weaker pulses. To further suppress the background noise, one can employ electronic gates that block signals below a specific threshold. Only pulses that are larger than a predetermined threshold are counted and are known as photon counting.

The recording and correction of fluorescence spectra requires good practice and is more difficult than the recording and correction of absorption spectra. Due to the high sensitivity of emission spectroscopy, it is possible to find and measure even the smallest concentrations of luminous substances, such benzopyrene, in river or industrial wastewater. On the other hand, contaminants with significant luminescence are detectable by emission spectroscopy of weakly luminous substances. These contaminants may be found in the sample, solvent, glassware used to prepare the solution, or even the sample cell. To account for minute impurity emissions as well as the solvent's Rayleigh scattering and Raman emission, a blank spectrum should be obtained with the pure solvent. A photoreaction of the sample that forms a fluorescent compound upon exposure to light or to the relatively powerful excitation beam of the spectrometer can taint the results, even after these trivial sources have been eliminated or at least characterized for later adjustment.

If Vavilov's rule is valid, excitation spectra are action spectra, and the emission intensity is relative to the quantity of light absorbed by the fluorescent sample at the excitation wavelength. Only when the excitation beam's intensity is calibrated by the quantum counter and the sample's absorbance is less than 0.05 will an excitation spectrum accurately mimic the fluorescent compound's absorption spectrum. Higher absorbances cause significant attenuation of light before it reaches the sample's center, which is focused on the monochromator slit.

3.11. MONOCHROMATOR

Slit width selection at the monochromators (Figure 3.12) is crucial. Better spectral resolution is achieved with smaller holes while less light is transmitted. Low light intensities can be made up for by raising the detector's sensitivity (higher voltage) and the integration times (slower scan speed). These, however, will also make the background more prominent. The slit widths should be selected independently, whereas the slit widths at the entry and exit of a monochromator should be the same for maximum performance. Short slit widths are chosen to enhance emission spectrum resolution, but small widths are necessary to produce an acceptable resolution of excitation spectrum. For analyzing mixtures, excitation spectra are extremely helpful. The excitation spectrum is likely to be significantly different from the absorption spectrum if the emission spectrum is caused by impurity. This can be used to determine whether an absorption spectrum created by irradiation is caused by a single photoproduct in low-temperature operationDiacetyl is a well-known example of an exception to the rule that few substances emit significant phosphorescence in the solution. Since phosphorescence span typically lasts milliseconds rather than nanoseconds, it is necessary to utilize highly purified solvents and thoroughly de-gas them to prevent quenching by dioxygen and other unintentional contaminants. A transparent quartz vessel filled with liquid nitrogen is used to submerge the sample cell, which is typically a quartz tube.

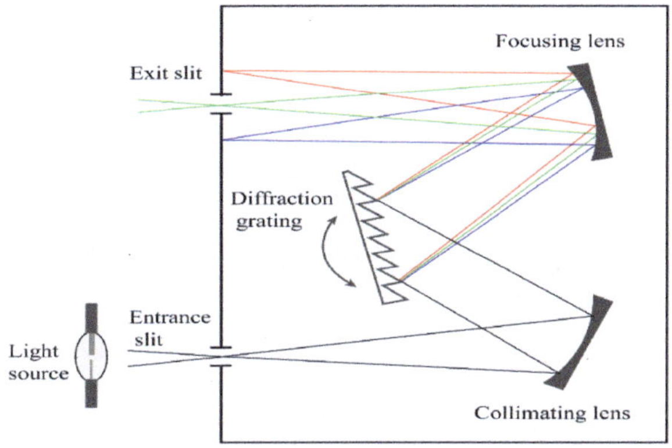

Figure 3.12. Monochromators.

Source: https://upload.wikimedia.org/wikipedia/commons/c/c6/Diagram_of_a_monochromator.png, Creative Commons License.

For phosphorescence studies of both organic and inorganic solutes, solutions that are solid at room temperature [borax or poly(methyl methacrylate, PMMA) glass] are an alternative. Warming methyl methacrylate solutions that have just undergone distillation for a few days causes them to polymerize in glass tubes. The finished mold may be easily cut to any desired form and polished to create transparent blocks. These samples can be used as reference samples to set up a phosphorescence spectrometer or for classroom demonstrations of the slow phosphorescence decay of naphthalene, for example, which lasts for many seconds. Since oxygen is primarily consumed during the polymerization process and air diffusion into the molds is relatively slow (1 mm per month), these samples are best used as reference samples. Even when solid sample solutions are investigated, phosphorescence spectra are frequently weaker than fluorescence spectra. Fluorescence must be reduced to detect phosphorescence. The sample is briefly exposed to the excitation beam by a hole in the cylinder, but the detector doesn't detect the sample during that time. Long after the fluorescence has dissipated, rotation of the container then allows access to the sample by the detector. The intensity dependence of the phosphorescence on rotational speed can then be used to calculate the phosphorescence span. In contemporary instruments, it is usual to eliminate fluorescence by switching to a pulsed light source, such as a stroboscope or a pulsed laser, and electronically deactivating the photomultiplier during and right after the excitation pulse.

Because the absorption coefficients for these prohibited transitions in organic molecules are so weak, singlet-triplet absorption spectra are challenging to detect. Although employing extremely high concentrations and path lengths is theoretically an option, in fact, this necessitates the use of materials with the highest levels of purity, and the technique fails at shorter excitation wavelengths when singlet-singlet absorption takes hold. With organic substances, few results have been attained in this manner. By employing heavy-atom media (CBr_4 or liquid xenon), and in particular, by paramagnetic enhancement under high pressures of dioxygen, the intensity of singlet-triplet absorption can be improved (oxygen pressure method). Evans conducted the majority of these tests in the 1960s using oxygen pressures of up to 130 atm. However, it is risky to expose solvents and reactive organic compounds to high oxygen pressure. The determination of singlet-triplet absorption spectra by monitoring the resulting phosphorescence emission has frequently been successful with comparatively little effort that because emission spectroscopy is much more sensitive than absorption, particularly for phosphorescence that can effectively be separated from light and fluorescence. Quantum mechanical simulations can forecast the direction of transition moments in the molecular framework. By observing the linearly polarized light that is absorbed by samples with specially oriented molecules, it is possible to experimentally determine the direction of transition moments. In single crystals, complete orientation is provided. However, optical measurements on single crystals are difficult because absorption studies call for exceedingly thin crystals.

CHAPTER

4

Crystallization of Organic Compounds

Contents

4.1. INTRODUCTION

The most vital separation and purification technique has been crystallization for decades. Due to several critical issues that necessitate improved control over the crystallization process, the study of crystallization processes has become even more important during the past few decades. These levels of control necessitate a deeper comprehension of both the principles and the operational traits of the crystallization apparatus. The need to reassure the regulatory authorities that a continuous supply of active pharmaceutical ingredients (APIs) (Figure 4.1) of high and reproducible quality and bioavailability can be delivered for formulation and ultimately to the patient reinforces the issue of better control (Elder, Snodin, & Teasdale, 2011).

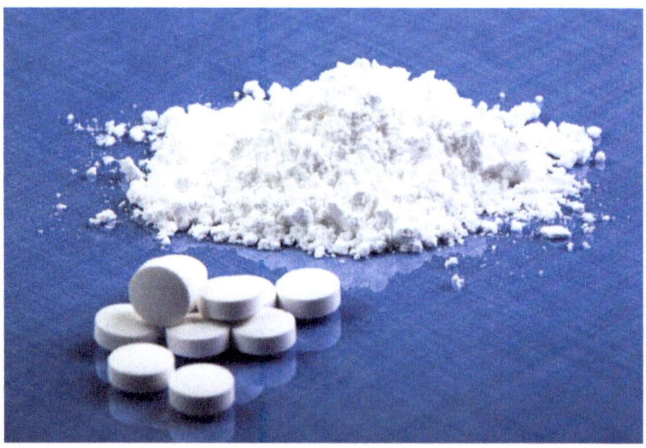

Figure 4.1. Active pharmaceutical ingredients.

Source: https://www.europeanpharmaceuticalreview.com/wp-content/uploads/APIs-e1613570763607.jpg, Creative Commons License.

The product image of the medication must match that employed in the clinical trials conducted to establish the product's legitimacy in the therapeutic market. According to the regulatory criteria, when scaling from pilot plant to industrial scale, physical properties like particle size distribution (PSD), bulk density, and surface area must be duplicated within specific limits. It is clear that crystallization has not been generalized to the same extent as distillation, extraction, adsorption, etc., as compared to the development of models and procedures for other unit processes. However, this is quickly changing as more research is being done on the principles of crystallization at academic and industry centers to model and forecast nucleation and/or growth rates as well as other important features, such as polymorph formation.

4.2. CRYSTALLIZATION

It is necessary to modulate either nucleation or growth or, as is most frequently the case, both types of crystal development simultaneously, in order to control the crystallization processes. From the perspective of the total result, each operation must be assessed to determine which of these process objectives is most important in order to decide whether nucleation or growth should be the dominant phase. For the obvious reason that the quantity and size of nuclei that initially form can dominate the rest of the process, a large portion of the literature is devoted to nucleation. However, it is commonly acknowledged that nucleation can be challenging to manage since a number of variables can affect the conditions in which nucleation begins, the rate at which it occurs, and the quantity of crystals that form before growth takes over. A change in focus from control of nucleation to control of growth has become necessary because of the requirement for more control over the physical properties of final bulk medicines. This trend is also being applied to downstream handling improvements and purity control for both intermediates and finished bulk products. The regulation of supersaturation and seeding then become the key variables. The development of a scalable process requires the quantification of these elements for each growth process.

In contrast to other crystallization methods like melt or supercritical crystallization, solution crystallization, in which solvents are utilized, is the predominant method of crystallization for pharmaceuticals and organic compounds. Different qualities have different levels of clarity. For instance, the amount of solid in equilibrium with the solvent is defined as solubility. The capacity of the crystallization process, as well as its capacity to reject undesirable compounds and limit loss in the fluid, can be affected by solubility. The solubility of a compound or solvent varies greatly from one to the next. On the other hand, other features are significantly less well understood or characterized. For instance, it is still unknown how oil or amorphous solids originate and under what circumstances. The composition of oil and amorphous solids might vary, and they can undoubtedly include a lot more contaminants than crystalline solids, which creates issues. Oil or amorphous solids are typically less stable and can provide serious problems for the stability of drug formulation and storage.

A crystalline compound's capacity to create polymorphs, more than one crystal shape for the same chemical entity is one of its characteristics. Because it influences all stages of drug development, from early drug discovery through final clinical evaluation, as well as patent protection and market competitiveness, the phenomenon of polymorphism is crucial to the pharmaceutical business. The early detection of probable polymorphs is a major challenge. Nucleation and growth kinetics must be carefully controlled and balanced to produce crystal products that meet specifications with a reliable, repeatable procedure. The fundamental mechanisms of nucleation and crystal formation, as well as the kinetics that follow, can be controlled by carefully controlling the environment. Unwanted polymorphs are frequently reduced to a minimum or removed by effective rate process management. Working with complex organic molecules can make it more difficult to manage crystallization processes in general. When trying to build a nucleation-dominated process, which, even under ideal conditions, can function over a very wide range of supersaturation, depending on tiny modifications like variable levels of very-low-level contaminants, this issue can become much more problematic. All of these issues have the potential to cause unintended solvent and/or impurity trapping in the final crystal. Even though many crystallization techniques can withstand a wide range of mixing quality and intensity, many researchers and developers do not consider the impact of mixing until scale-up issues or perhaps even at the laboratory scale compel them to.

When it comes to complex organic molecules in particular, the transport of momentum, mass, and energy are impacted by mixing and can be crucial to the success of many crystallization procedures. Homogeneity, impact nucleation, shear damage, agglomeration, and discharge can all be impacted by momentum transmission. Mass transfer can have an impact on the consistency of supersaturation and reactive crystallization can have an impact on the subsequent reaction and supersaturation pattern even at the molecular level. Heat transfer is directly influenced by the flow of energy, and effective mixing can reduce or even prevent encrustation on the surfaces of heat transfer. The Damkoehler number (Da) in its modified form is a helpful concept for assessing mixing effects during crystallization (Stenberg, Stiblert, & Nygren, 1986). These processes employ seeding to variable degrees and supersaturation in diverse ways. The methods are categorized according to how supersaturation is produced. Each classification introduces the equally crucial questions of when to seed and how much seed to utilize. The amount of seed can range from none to massive, and it can be categorized into the well-known categories of a pinch (to prevent complete nucleation), small (1%), large (5–10%), and massive (the seed is the product in a continuous or semi-continuous operation) to maximize growth. In the control of polymorphs and hydration/solvation, the amount of seed can also play a crucial role.

4.3. TOOLS

Powerful tools are being added to help with the control of crystallization activities in both laboratory and manufacturing operations by developing methods of online assessment of solution concentration and particle size and count. Depending on the system and the importance of the desired outcome, there are numerous techniques to cool a solution from its solubility temperature. The simplest way is natural cooling, which is dependent on the crystallizer's ability to transfer heat, although it causes supersaturation as cooling progresses. Depending on the specific system's nucleation and growth rate characteristics, this might or might not be bad for the process. Before seeding may be effective, natural cooling may be able to bring the temperature down quickly enough to pass through the metastable phase and reach the uncontrolled nucleation region. Uncontrolled nucleation has the potential to lead to oiling out, fine particle agglomeration, greater PSD, and the occlusion of solvent and contaminants, among other serious issues. Crystal scale buildup on the cooling surface brought on by low temperatures at the wall is a secondary drawback of unregulated cooling. Nucleation on the cold surface followed by development might cause the accumulation of a scale layer. This encrustation can significantly slow down cooling and result in significant product nonuniformity problems. Cooling techniques can be used to match the cooling rate with the growing surface area when high supersaturation is unacceptable.

4.4. SURFACE

When the crystal surface area for growth is low, we can use cooling rates that are much slower initially than natural cooling to preserve supersaturation in or around the growth region. As the surface area grows, the cooling rate can rise. This technique also can lessen encrustation by reducing temperature variations throughout the sheath. Theoretically, if the temperature difference between the cooling fluid and the crystallizing mixture is less than the diameter of the metastable zone, encrustation can be prevented. Researchers reported a further improvement of this method in which a seed age is introduced as an aid in restricting the expansion of supersaturation, hence minimizing nucleation, and encouraging growth. Seeding is another important factor in batch cooling. The point at which the batch temperature first crosses the saturation curve is difficult. However, batch-to-batch fluctuations in several variables, such as the precise concentration of the substance to crystallize and contaminants that may impair solubility, can have an impact on this temperature. Some or all of the seed may dissolve if it is added at a temperature higher than its solubility temperature, which could lead to uncontrolled nucleation. The product can already have nucleated if the seed is applied at a temperature that is too low for saturation. In either situation, a rise in nucleation might lead to a decline in impurity rejection as well as/or a modification of PSD and other physical characteristics. Finding the point of seeding is a problem that arises during solvent removal by concentration, antisolvent addition, and cooling-induced crystallization.

A growing number of crystallization operations are using online, in-situ instrumentation to assess product composition to successfully determine the seed point. The existence of nuclei can be detected via image analysis or photography techniques. It would be too late to establish the point of seeding at 5 microns. However, these techniques can be used to assess whether seeding was effective and to check for excessive nucleation. The nucleation/growth ratio can also be brought within the typical range by including an aging time at constant temperature after seeding. One of the finest ways to regulate batch-cooled crystallizer is to add the seed as a slurry in the appropriate solvent composition. Before saturation, the slurry is added, until it is clear that the seed is no longer dissolving. The amount of seed that remains may be prone to excessive variance, even though this procedure can improve the likelihood that seed will be present at the beginning of crystallization. When polymorphs are stable at various temperatures within the cooling range, crystallization by cooling may not be possible. A transition from one polymorph to another or mixed morphologies might arise from cooling through these stable areas. When polymorphs are possible, uncontrolled nucleation can also pose a significant challenge to producing a consistent result. To choose the appropriate polymorph, a constant-temperature method with either a high level of seed or enormous seed may be necessary. These elements may also affect hydrates and solvates during the crystallization process. It is possible to regulate low degrees of supersaturation and operate with huge amounts of seed by using continuous operation, which largely eliminates the batch-to-batch fluctuation for batch cooling systems.

Both crystallizers' seed beds are large in comparison to the volume of racemic solution flowing through them to provide enough seed area and maintain low supersaturation. In the overhead streams, uncrystallized isomers are recycled to dissolve more racemic feed. Sonication helps to retain crystal size. Because the resolution is not reliant on equilibrium solubility, batch cooling is not a workable alternative to achieve the separation of optical isomers. When any practical amount of product is to be harvested in each cycle, the time needed for batch cooling would lead to the nucleation of the undesirable isomer. In constantly stirred tank crystallizers, a high level of control is also possible. Temperature variations between the feed and the crystallizer can be controlled as needed. The feed rate, concentration, and solubility differences attained will determine the seed's presence at the slurry concentration. Cross-flow filtering on the discharge of the crystallizers can be employed to improve the slurry density in situations where the amount of seed is insufficient. Semibatch operations, which increase concentration by evaporating solvent, are frequently used but have several nucleation and growth control issues. In some circumstances, such as for final bulk drug compounds that may require tighter control of mean particle size and PSD than can be obtained on scale-up, these issues can be so severe as to render this procedure inappropriate.

4.5. SATURATION

In achieving supersaturation, evaporation rate is parallel to cooling rate, and both can be regulated using methods that match evaporation rate with surface area available for growth. Since it might be challenging to tell when the saturation line is being crossed as concentration rises, the point of seeding is another problem. In this case, adding the seed as a slurry to the evaporation solvent as the concentration approaches saturation can be helpful. The most important control problem that can lead to solvent and impurity occlusion, non-reproducibility in PSD, and other physical properties is local fluctuation in supersaturation. Both the heating surface and the boiling liquid/vapor interface experience these differences. Locally high temperatures and rapid vaporization at the heating surface produce unmanageable local supersaturation situations where excessive uncontrolled nucleation may occur, especially in areas with poor bulk mixing. Important concerns with product quality might also result from wall scale above the heated surface. Because of the direct contact with the higher temperature of the heating fluid, decomposition on the surface above the liquid-vapor boundary may be excessive. Product scale from this location may also fall into the product slurry and cause handling issues in any system as well as undesirable physical qualities for the final bulk drug. If the concentrated mass is thermally unstable, overconcentration may cause safety problems.

It is a potential negative effect of an evaporative crystallization operation even though it is not a crystallization issue. Vapor near the boiling surface can result in very high local supersaturation as well as nucleation brought on by vapor-liquid interfaces. Foaming can also be a serious problem. Vapor bubbles in the bulk can also result in local nucleation. All of these causes of variability contribute to serious scale-up issues. Higher wall temperatures may be necessary to manage the distillation rate by adjusting the jacket temperature, which would exacerbate supersaturation variation. This issue will be made worse by the reduction in bulk circulation and the expansion of the mixing process. These issues occasionally lead to unsatisfactory results, necessitating the creation of a different crystallization technique. It is frequently not possible to operate continuously to provide the conditions for crystallization. Uncontrolled nucleation is caused by local supersaturation at the liquid-vapor-solid interfaces. In terms of crystallization control, this commonly utilized method has numerous advantages. It has the apparent drawback of producing solvent combinations that need to be separated for recovery. Controlling the antisolvent addition rate makes it simple to manage both supersaturation and the area where crystals develop (Duffy, Barrett, & Glennon, 2013). This control is possibly more complex than the cooling and concentration procedures for single-solvent systems because it takes into account both the change in solubility as addition progresses and the crystal development region. Using a sizable chunk of the prior batch as the seed also enables massive seeding.

When scaling up these procedures, it is important to pay close attention to how the antisolvent is mixed both at the point of addition and during bulk circulation. Local supersaturation and excessive nucleation can occur as a result of insufficient management of local mixing at the point of addition. Subsurface addition of the antisolvent can, in some situations, be absolutely necessary for successful scale-up in order to reduce this risk. Overmixing is a problem as well since shear can split crystals and cause secondary nucleation, which produces nuclei. Loop design was created for reactive crystallization and can also help antisolvent crystallization methods be scaled up. Semicontinuous techniques, where the supersaturation is regulated locally at the site of mixing in an in-line device, can provide excellent control of the crystallization conditions. An in-line device can, in the case of unseeded operation, quickly produce a high supersaturation ratio and offer a way of nucleation control that is difficult or impossible to achieve in conventional crystallization vessels. With the right design, mixing at the molecular level can be completed faster than nucleation, leading to a process that produces homogenous, tiny particles largely by nucleation. Following the nuclei's exit from the mixing zone, further crystallization occurs in an agitated vessel on a predetermined initial batch size and form of nuclei. This method is being used to produce bulk drugs because it can produce narrow PSDs with regulated surface areas. The ability to manage particle size also has the added benefit of removing the requirement for milling to reduce and control particle size. By operating at the same local conditions in the same size jets for longer periods of time, scale-up to production scale can be accomplished.

4.6. REACTIVE CRYSTALLIZATION

Reactive crystallization (Figure 4.2) is the process by which a substance that crystallizes becomes supersaturated as a result of a chemical reaction (McDonald et al., 2021). A complex chemical may combine with another complex organic component to generate a salt, or a complex compound may be neutralized by an acid or base. These reactions may occur very quickly in comparison to the crystals' growth and mass transfer rates, which can result in local supersaturation and nucleation. Because of the quick intrinsic kinetics, these activities are also known as precipitations. It is challenging to regulate particle size in reactive crystallization because, in most cases, there is no practical way to slow down the process that causes the supersaturation. However, the rate at which the chemicals are added does offer a way to manage this important parameter in the reactor because the reaction can be finished just where the reagents are added. Therefore, a careful balance must be struck between the rate of reagent addition, mass transfer, and crystal formation surface area. To avoid uncontrolled nucleation and the subsequent production of an excessive number of particles, controlled supersaturation at the start of the addition of the reagents requires an initial charge of seed. The intrinsic reaction may only produce crystals that are too minute to be utilized as seed if a simple growth process is required, hence the seed must be created in a separate operation. In some situations, the use of ultrasound in crystallization can be unusual and quite beneficial. Ultrasound was utilized to break apart crystals and create a new surface area for subsequent crystal formation. Increasing the crystal aspect ratio by fragmenting crystals along the long axis and/or by promoting development on the slowest-growing surface can be highly helpful.

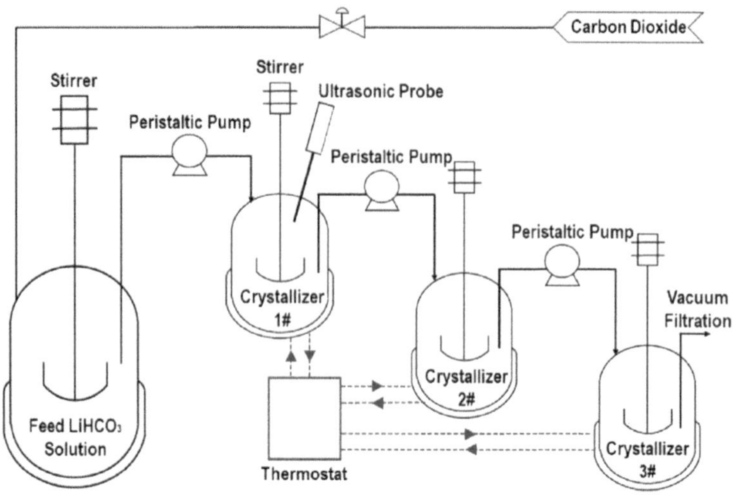

Figure 4.2. Reactive crystallization.

Source: https://www.researchgate.net/publication/353663975/figure/fig1/AS:1057840030576640@1629220125874/Schematic-image-of-a-three-stage-continuous-reactive-crystallization-experiment-setup.png, Creative Commons License.

The production of nanoparticles to increase the rate of drug dissolution is one major factor driving supercritical crystallization. This strategy shares many fundamental similarities with impinging jet crystallization where high supersaturation is produced by quickly combining two streams. Supercritical crystallization operations of various types have been successfully developed for multi-kilogram scales. Now that online measurement techniques like focused beam reflectance measurement (FBRM) and Fourier transform infrared (FTIR) have advanced, it is possible to quickly get information on the PSD and solution concentration using these in-situ probes. Hundreds of data points can be produced in a single experiment (Yingying, Bao, Wang, & Rohani, 2012). The model-based experimental design for crystallization can produce high-quality crystallization kinetic data from a limited number of tests with careful experiment design. Thus, this method can significantly reduce the amount of time and experimental work required to create crystallization techniques. More mixing systems are being analyzed using computational fluid dynamics (CFD), particularly the stirred vessels frequently employed for crystallizer operation. Although modeling fluid dynamics in the presence of a solid phase is a challenging task, some researchers are beginning to make progress in this area. The development of the process as well as the production operations would have benefited from the use of these methodologies. There is a large body of research outlining these techniques, such as feedback control of supersaturation for crystallization, and the instrumentation used for measurements and control is continually getting better.

In addition to internal standards, such as consistency for intermediates and especially for active APIs, controlled crystallization techniques and apparatus are necessary to satisfy regulatory requirements. Controls on both chemical purity and physical characteristics are part of these standards. The bio batch model for clinical evaluation sets constraints on chemical purity, mean particle size, PSD, and other suitable physical properties for APIs. The term "bio batch" refers to the legal requirement of identifying a specific batch as the defining standard for physical and chemical attributes that must be reproduced at the industrial scale to be acceptable for sale. This particular batch is a pilot scale batch used in clinical trials. Once defined, the critical process attributes (CPAs) must be met while scaling up to the manufacturing facility. In order to develop a crystallization process, it is necessary to identify achievable and repeatable ranges for both the CPPs and the CPAs. In contrast to other crystallization methods like supercritical crystallization, solution crystallization, which employs solvents, is the main crystallization approach for pharmaceuticals and fine organic compounds. Based on our own experience, we will concentrate on the specific features that are more pertinent to solution crystallization and those that directly affect the quality of the final bulk drugs, such as purity, form, and size. It is essential for the successful development of crystallization methods to comprehend the solubility behavior. To carry out the crystallization process, it is required to know how much solute can dissolve in the solvent system initially and how much will remain in the solvent at the end. Chemical solubility for solution crystallization is simply the maximum quantity of the compound that may dissolve in a given solvent system. If the solute concentration reaches the solubility limit, the solution is said to be saturated. When a solution reaches saturation, no further dissolution will take place, and the amount of dissolved solute in the solution stays constant. As a result, in this situation, the solid and dissolved solutes are in equilibrium.

At higher temperatures, solubility in some substances, particularly inorganic salts, may not be altered or even decline. Hard Water, which contains calcium carbonate, is a better example because it dissolves slowly at higher temperatures. Because water is kept at a higher temperature in the water boiler, this reverse solubility behavior leads to a scale deposit problem. Calcium carbonate consequently experiences supersaturation and precipitates in the boiler. Reverse solubility, however, is uncommon in fine organics or pharmaceuticals. Temperature is frequently used to regulate the crystallization process since it can have a big impact on solubility. When the temperature is lowered, the pure material can crystallize from the solution whereas the impure material can simply dissolve in a specific solvent system at a higher temperature. In some solvent systems, solubility behavior can be nonlinear. Acetonitrile serves as an antisolvent. The solubility curve increases at the middle of the solution mixture and decreases with higher or lower acetonitrile concentrations. The occurrence of azeotrope, which has a lower boiling point (BP) than pure component, would be an analog of nonmonotonic behavior to vapor-liquid equilibrium. Solvent composition is a variable in regulating the crystallization process, similar to temperature. The pure substance can be crystallized from the solution by adding the antisolvent after the impure material has been dissolved in a specific solvent system. Other crystallization factors, such as polymorphism, solvate, crystal shape, crystallization kinetics, etc., can be significantly influenced by solvent.

4.7. CONTAMINATION

The effect of contaminants on solubility in practical applications is typically unknown a priority and must be evaluated experimentally. If the starting materials have different levels of impurities from batch to batch, caution should be used when conducting crystallization investigations due to the potential impact of contaminants on solubility. Crystallization kinetics can also be impacted by the presence of impurities. We typically presume that two substances will have similar solubilities if they have similar chemical structures. Simvastatin has one additional methyl group, however, the solubilities of the two compounds differ substantially. A compound's solubility can be changed by the formation of salt if it contains one or more acidic or basic functional groups. This is yet another helpful method for carrying out crystallization because altering the salt form can have a major impact on solubility. For instance, the intended chemical might not be very soluble in the selected solvent. However, once the salt has formed, its solubility can substantially rise and become soluble. The needed pure salt can be crystallized by cooling or by adding an antisolvent. As an alternative, the desired compound would dissolve in the chosen solvent and change into a salt that is less soluble in it. As a result, the salt will form crystals from the solution, a process known as reactive crystallization.

Similar to how changing the solvent impacts other crystal features including shape, polymorphs, crystallization kinetics, formulation, drug stability, etc., changing the salt form has a similar effect. As a result, in addition to solubility, other factors are always taken into account while choosing the salt form. Individual solubility measurements can be made in the lab readily with a simple setup. An excessive amount of substance is charged into a jacketed glass flask along with a solvent or solvent mixture. The slurry is stirred and kept at the appropriate temperature for a specific time, ideally overnight. The slurry is then sampled and run through a filter. For a high-performance liquid chromatography (HPLC) experiment, the clear filtrate is diluted. For simplicity of handling, sampling, and filtering of the slurry are typically done at room temperature. If the temperature at which solubility is tested is different from the ambient temperature, this may result in measurement errors. There are numerous ways this technique could be altered. For instance, a filter can be added to the glass flask used for measuring solubility in order to do in-situ filtration. High-throughput screening tools, which were first used in the combinatorial approach to drug discovery, are now being used frequently due to the limited material supply in the early stages of drug development and the requirement to quickly determine the proper solvent and temperature range for crystallization. The high-throughput screening equipment often has many cells or vials ranging in size from microliter to milliliter, allowing users to run several experiments with a relatively small amount of materials. Automation is the main technological achievement in this field since it allows for the repetitive handling of several samples. The use of in-situ analytical instruments, which are a subset of process analytical technologies (PATs), such as FTIR, Raman, ultraviolet/visible (UV/Vis), or near-infrared (NIR) spectrophotometers, is another trend in the industry. These devices frequently have a sensor probe that may be dipped into the batch or process stream. PAT is more suited to process monitoring than analytical analysis, as opposed to the high-throughput screening method.

4.8. SPECTROPHOTOMETER

Spectrophotometers (Figure 4.3) can be used to gauge a solution concentration or solubility. Calibration with known concentrations is necessary to determine the relationship between solubility/solution concentration and spectral peak regions. In-situ analytical instrumentation does have some restrictions and concerns, though, such as minimum detection level, peak shift, peak overlapping, etc. It can still take some work to create an accurate correlation model for a given molecule across a broad range of temperature and solvent composition. Solubility prediction is getting more attention, mostly from academic studies. The prediction of solid-liquid equilibrium is still in its early stage in contrast to the vapor-liquid equilibrium condition. But given how quickly this sector is growing, one cannot ignore its future potential. When the solute concentration surpasses its solubility limit, the solution is supersaturated. For a specific time, a solution may remain supersaturated over a concentration range without the development of a secondary phase. The period between the onset of supersaturation and the initial appearance of the secondary (solid) phase is known as the induction time. The induction time decreases with increasing supersaturation. When supersaturation reaches a particular point, the secondary phase starts to emerge on its own as soon as supersaturation is produced. The metastable zone width is referred to as this location. Similar to solubility, the temperature, solvent composition, chemical structure, salt form, contaminants in the solution, and others have an impact on a supersaturated solution's metastable zone width and induction time. Consequently, even though the spinodal point is a thermodynamic feature, it is highly challenging to determine the absolute value of the metastable zone width empirically. Designing crystallization methods can benefit from an understanding of the qualitative behavior of the metastable zone width and the induction time.

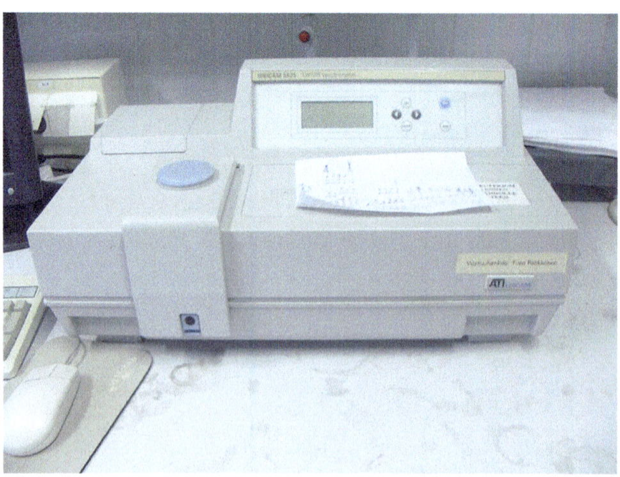

Figure 4.3. Spectrophotometer.

Source: https://upload.wikimedia.org/wikipedia/commons/thumb/1/10/Spektrofotometri.jpg/800px-Spektrofotometri.jpg, Creative Commons License.

If there are no extra problems, such as thermal or chemical decomposition, it is reasonable to predict that the distance between the spinodal and binodal points should decrease as the temperature of the system increases from the free energy-composition phase diagram. This anticipation is predicated on the idea that two phases will mix readily when temperatures are high enough. It appears reasonable to anticipate that the metastable zone width and induction time will be shorter for a lower antisolvent level for reactive crystallization. However, more information is required to support this idea. There could be a lot of exceptions to this theory. Regarding contaminants, the metastable zone and induction time often grow wider and longer as the concentration of impurities in the solution rises. The metastable zone width and induction duration can also be influenced by other elements, including the presence of desired compound seed, extraneous solid particles that cannot be dissolved, and even the degree of agitation. These elements undoubtedly can disturb and change the free energy-composition phase curve. The metastable zone width and the induction duration will both often decrease as a result of these considerations.

4.9. SOLUBILITY

Similar to measuring solubility, it is possible to determine the solution concentration offline, whether in supersaturated or saturated states, by taking slurry samples from the process. Due to temperature variations and solvent evaporation, this approach has drawbacks. Sampling typically requires a lot of time and efforts. As previously indicated, given the solubility information, in-situ analytical tools like FTIR or UV-visible spectrophotometry can assess the solution concentration and calculate the supersaturation. The concept of crystallization kinetics and the progress of the crystallization process can both be improved by accurate supersaturation detection. It takes more time and effort to accurately determine the metastable zone width and induction time than it does to accurately determine the supersaturation. Therefore, if the working environment is altered, the data might not be repeatable. On the other hand, many industrial applications do not require precise determination of the metastable zone width or the induction duration. For the design of crystallization processes, a few data points for preliminary estimation are adequate. The first step in measuring the metastable zone width is to make an undersaturated solution based on solubility information. The transparent undersaturated solution is cooled at a defined rate to produce supersaturation to cool crystallization. The boundary of the metastable stable zone is shown during cooling by the solution becoming foggy at the specific cooling rate. This method may need to be repeated numerous times with various cooling rates because the measured metastable zone width can vary depending on the cooling rate. The true metastable point is defined as the asymptotic point with a cooling rate of zero. This procedure will be repeated at various initial solution concentrations to create the whole metastable zone width curve. Similar methods can be created to measure the metastable zone width for antisolvent crystallization. Using an in-situ particle size analyzer, such as the FBRM equipment, which can detect and quantify the particle size without a sample for external examination, can simplify the process of determining the metastable zone width (Kim & Koo, 2012).

The explanation of solubility above demonstrates that, if such a curve is available, it is theoretically possible to determine the metastable zone width through the free energy-composition curve. For practical applications, experimental verification is recommended due to the complexity of the issue. In general, the operational window for supersaturation is lower for batch crystallization with a narrower metastable zone width. With fine crystals, nucleation is more likely to occur, and vice versa. However, having a wide metastable zone width is not always preferable. L-tartaric acid was added to a free base solution in an acetonitrile/ethyl acetate/water solvent mixture to create the tartrate salt. If the seed surface area is not high enough, the solution stays strongly supersaturated (metastable) for an extended period after the addition of tartaric acid to the free base. Problematic conditions include the presence of a highly supersaturated metastable solution. First, the batch time cycle will be greatly extended by the delayed release of supersaturation. The extremely supersaturated solution may also cause unpredictable nucleation, which would be challenging to control and result in variable performance from lab to factory and batch to batch. The dissolved substance will separate from the solution upon release of supersaturation and produce a secondary phase, which may be oil, an amorphous solid, or a crystalline solid. Molecules are organized in solids called crystalline materials in three-dimensional patterns. Solids called amorphous materials lack a regular three-dimensional pattern in their molecules. In some cases, with extremely high supersaturation, the initial secondary phase may be a liquid phase, such as oil, where molecules may be randomly organized in three-dimensional patterns and exhibit significantly better mobility than solids. In general, the oil phase is unstable and will eventually change into an amorphous substance and/or a crystalline solid. An amorphous solid can be produced at a lower level of supersaturation. The amorphous solid, like oil, is unstable and can change into a crystalline solid over time. There could be various solid states with various crystal shapes and levels of stability, even as a crystalline solid.

4.10. CRYSTALLIZATION

During the crystallization process, materials can take on the states of oil, amorphous, and crystal. Differential scanning calorimetry is an often-employed method. The heat flow profile over temperature that results from heating a solid sample at a predetermined heating rate displays how crystalline the solid sample is. To detect an endothermic heat flow that corresponds to the enthalpy of melting at the material's melting point, the heat flow profile should be smooth during the whole heating period. If there is any amorphous solid material in the solid, there will likely be some solid-state transition and associated heat flow before melting. It is demonstrated in the study of simvastatin that differential scanning calorimetry can be used to distinguish the flaws of crystals produced by various crystallization techniques. Oil droplets typically have extremely high impurity concentrations, including solvents, and provide purification challenges. The oil phase cannot be separated as a solid for use in drug formulation experiments because it is essentially a liquid phase. With little control over the particle size, the oil phase is unstable and will probably phase into an amorphous or crystalline solid. As a result, the process suffers from the creation of the oil phase. To reduce this problem, common strategies include seeding and supersaturation control. Similar to oil, amorphous materials have problems with impurity and solid-state instability. Even though occasionally the compounds are unable to produce a crystal structure, the creation of an amorphous solid is normally undesirable. Some drugs, especially those made by freeze drying, are naturally amorphous solids (Yu, 2001). The primary goal of these amorphous drugs is to increase the pace of drug degradation in the body but at the expense of drug stability and sub-ambient storage temperature. Though amorphous solid dispersion has recently received increasing attention, crystalline solids are still by far the most ideal form for drug development. In addition to crystallization, maintaining crystallinity is necessary during processes like filtration, drying, and milling. A hygroscopic solid may lose its crystallinity after absorbing water. Controlling the humidity during drying and storage is therefore essential for this kind of material. With drying or milling, some crystalline materials may tend to lose their crystallinity. Crystals with polymorphism display a variety of physical characteristics and behaviors, such as solubilities and melting points (MPs). Beyond crystallization, the significance of identifying the drug's stable polymorphs in the drug development process cannot be overstated. All past developmental results, including the crystallization of pure bulk drugs, need to be re-evaluated if a more stable form is found in the late stages of drug development. For the same reason, the effects may be significantly severe if the stable version is discovered after the medicine has been sold. Fortunately, the latter situation is not common.

A thorough understanding of the behaviors of solid-liquid equilibrium under various situations, such as temperature or solvent mixture, is essential. The expected polymorphs must be maintained during the entire operating sequence, which includes filtration, drying, and milling. If the drying temperature is higher than the crossover temperature, the enantiotropic system may undergo switching from one polymorphism form that is stable at room temperature to another polymorphic form that is stable at a higher temperature. Keeping the drying temperature below the crossover temperature would be ideal. Essentially, the measurement of solvates and non-solvates can be done using any of the analytical instruments. A solid's mass can be analyzed using gas chromatography (GC) to establish the presence of solvent in a solvate. To prevent the transition from solvate to non-solvate in the solid form, special care should be taken when preparing the sample. The wet cake solvate frequently loses its solvent while drying. The dry cake crystallizes into a non-solvate solid, which is not the actual solvate solid. A new substance known as a solid compound is produced when a compound adds another substance besides solvent to its crystal lattice. The ratio of the 1st compound to the 2nd compound in the solid crystal lattice is set, just like with solvates. The ratio can, however, change dramatically under some circumstances. In this case, the new substance is referred to as a solid solution. This indicates that the two compounds can freely mix to form a new compound, much like two miscible liquids. A solid mixture is a combination of two substances that have been physically blended. Each component still has its crystals when viewed in detail. No novel crystal structures or compounds are created.

To break solid compounds or mixtures for the optical isomers, an optically active resolving agent is usually added. The resolving agent can either be acidic or basic, depending on the chemical composition of the optical isomer. The solid compound in its initial neutral state can be changed by the resolving agent into a solid mixture in its salt form by a reaction with the optical isomers. Following that, the solid mixture can be divided using the solubility difference. Crystallization can be used to kinetically separate the diastereomer salts from one another if the equilibrium solubility difference between the two salts is insufficient to produce a significant yield. Ibuprofen-lysine diastereomer salt resolution is an example of this (Fogassy et al., 2006). When impurities or solvents are physically trapped within crystals, inclusion or occlusion occurs. In contrast to solid compounds, where impurities are dispersed across the crystal lattice, inclusion or occlusion allows impurities to be intermittently occluded within crystals. While both inclusion and occlusion enclose solvent or impurities in the crystals, inclusion is more applicable to situations in which surface liquid is trapped within crystal clusters or agglomerates during drying, whereas occlusion is more applicable to situations in which solvent or impurities are trapped within crystal cavities during crystallization. As opposed to direct microscopic inspection of the occlusion of crystals after drying, the presence of occlusion is generally inferred from the residual solvent profile during drying. The wet cake can be thought of as a matrix of crystal solids. Crystal solids can have a thin layer of solvent applied to both their exterior surface and internal holes. The external solvent will evaporate during drying, which may cause the dissolved solute to crystallize on the surface. The external surface of crystal solids can be sealed off if there is a large amount of crystallized material. This traps the solvent in the interior space. With prolonged drying, the solvent cannot be removed anymore. The antisolvent may also evaporate before the solvent when a mixed solvent is dried. Agglomerates of dry solids are frequently seen, and despite prolonged drying, the cake's residual solvent level stays constant. To limit the amount of dissolved solids in the solvent on the crystal surface, it is helpful to keep the drying temperature below specified values.

4.11. ANALYSIS

The solvent vapor pressure at the drying temperature must be kept sufficiently below the dryer pressure. This guarantees efficient solvent evaporation from the wet cake's surface. Two widely utilized procedures for the calculation of how much impurity and leftover solvent is in the cake are HPLC and GC analyses. Another extremely effective tool is thermal gravimetric analysis. It measures the temperature at which the solvent evaporates in addition to the amount of residual solvent present. Solvent is trapped inside the cake if the weight loss of the cake from solvent evaporation occurs at the solid's melting point. HPLC can be used to measure. To distinguish between potential mechanisms during crystallization, it gives useful information on the trends and distribution of impurities in the dry cake based on the total material balance in the cake. Solvent and temperature can have a significant impact on crystallization behaviors, making them important factors. Therefore, it is preferable to systematically screen the solvent and identify the ideal crystallization conditions. It is expected that there will be a significant advancement in this area future shortly given the quick development of high-throughput screening tools for the assessment of solubility. The choice of the supersaturation generating method and crystallization mode for solvent systems with a window of operating temperature can also have an impact on the overall crystal growth rate (Zhang, Szilágyi, Gong, & Nagy, 2019). Adequate mixing to prevent local excessive supersaturation can be crucial in many situations when solvent or impurity rejection becomes crucial.

A completely different mechanism can be used to bind solvents or impurities to crystals. Crystals' surfaces are capable of adsorbing solvents or contaminants from the surrounding liquid or gas phases. As a result, contaminants or solvents can move between the liquid/gas and solid phases. Different adsorption mechanisms have been postulated depending on the type of interaction. By comparing the impurity levels in the solution before and after adding the solid, the adsorption isotherm can be constructed. The adsorption isotherm can then be identified based on the alteration in impurity concentrations in the solution. In the specific instance of hygroscopicity, water is preferentially adsorbed onto the surface of the crystal. We refer to substances that display this tendency as hygroscopic substances. A hygroscopic substance can liquefy by a process known as deliquescence in which a small amount of water is absorbed by the crystal surface. Pharmaceuticals frequently contain hygroscopicity and deliquescence, which decrease the effectiveness of impurity rejection through crystallization if contaminants are preferentially absorbed on the crystal surface. If not carefully handled, trace amounts of adsorbed contaminants may have negative effects on the resolution of chiral molecules or the crystallization of pure bulk drugs. To get rid of the contaminants that have been adsorbed, an efficient wash is highly desired.

4.12. SOLVENT

Adsorbing solvent won't hinder processing because it may be washed away after drying. If the solvent in the wet cake becomes trapped inside crystals when drying, as was discussed in the preceding section, complications may result. The interaction of the solvent mixture on the crystal surface when a solvent mixture is used to crystallize is influenced by the affinity to crystals and can be complex. Hygroscopic material needs to be handled and processed carefully. It is crucial to preserve a dry atmosphere during drying, packaging, and storing to remove the water and keep the cake dry. Moisture from the vapor phase will be absorbed by the cake. A wet cake might not meet the required standards or have poor long-term chemical stability (Georg, Reynolds, Frank, & Halliday, 2006). When exposed to a moist environment, some materials may deliquesce, undergoing an irreversible transformation from a solid state to an amorphous one. When drying, packaging, and storing hydrates, it is preferable to maintain a specific relative humidity level. The formulation may dry out and dehydrate if the humidity falls below a particular point. Crystal forms are frequently observed with optical and scanning electron microscopes. To measure offline, samples from the crystallizer might be taken. Therefore, it is important to take precautions to prevent sampling-related problems. Commercially available equipment for in-situ observation without sampling is also available. Due to the optical focal depth limitations, these devices can only resolve objects within tens of microns. The morphology of crystals can now theoretically be predicted.

The development of theories has greatly improved our comprehension of how molecules are oriented both inside and externally within crystal lattices. As a result, instances in which the relative growth rates of various surfaces can be altered with additives or solvents have been documented. The development of crystallization processes benefits from an understanding of the morphologies of crystals based on varying crystal growth rates on various surfaces. Generally speaking, the surface area for development is substantially smaller in needle-like crystals. To offer an appropriate surface area for crystal formation, additional seed is therefore needed. On the other hand, crystals that resemble rods or cubes have a lot more surface area where they can grow. Therefore, less seed is needed. The fastest-growing surface frequently outgrows itself and disappears. As a result, beyond a certain stage of crystallization, the apparent release of supersaturation by crystal development is significantly reduced. If it happens during crystallization, the supersaturation will increase and there may be a brief spike of nucleation. Crystals with needle-like and plate-like shapes complicate the procedure even further. For instance, compared to cube-like crystals, these crystals often have higher filtering resistance and worse solid flow properties for formulation. Therefore, growing bigger crystals is extremely desirable. To develop thick crystals, it is important to conduct experiments to determine the ideal solvent for doing so. Meanwhile, in various solvent conditions, solvates and hydrates can develop. It is also possible to assess chemical forms like salt, free base, and free acid.

4.13. SONICATION

Sonication (Figure 4.4) is another option for modifying crystals that resemble needles. One crystal is split into two or more parts along its longest dimension by sonication (Chattopadhyay & Gupta, 2001). As a result, the crystal's length or apparent growth rate in this dimension is effectively decreased.

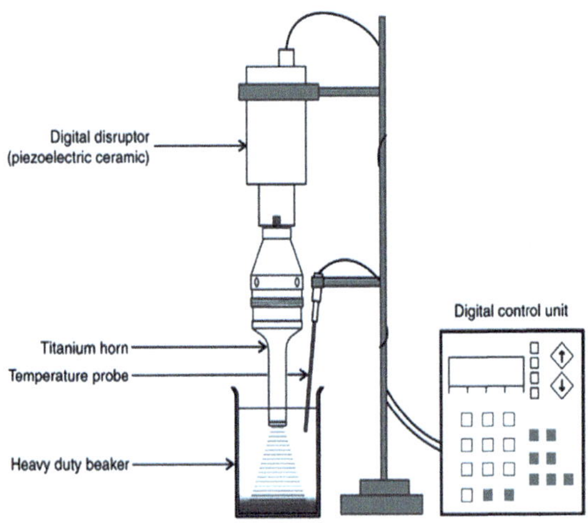

Figure 4.4. Sonication.

Source: https://ars.els-cdn.com/content/image/3-s2.0-B9780128195406000012-f01-07-9780128195406.jpg, Creative Commons License.

The crystal's shorter dimensions, however, are unaffected by sonication and continue to develop. As a result, the crystals gradually become thicker and shorter. Another special characteristic of crystals is their PSD. Crystals come in different sizes and forms. The PSD function is a very useful tool for describing how crystals of various sizes distribute their particle sizes. Despite the three-dimensional length of crystals, the one-dimensional PSD function is widely employed in real-world applications. It might be more practical to extend the one-dimensional PSD function to two or three-dimensional PSD functions. However, these functions are undoubtedly more complicated, and it is not always clear how they are better than the one-dimensional PSD function. PSD measurement has been done in a variety of ways. Sieving, laser scattering, sedimentation, image analysis, and other techniques fall within this category. Laser scattering devices, such as HELOS or Microtrac (forward scattering devices) or FBRM (backscattering devices), are one of these techniques that are used in industry (Roostaei et al., 2020). There are various tools for measuring dry and wet systems online. Various types of instruments will probably produce different PSDs for nonspherical particles due to the diversity of measuring methods and the differing definitions of characteristic length. As a result, it is important to realize that there is no definitive standard for measuring PSD.

Understanding the PSD during crystallization can help us better understand how the system behaves during crystal nucleation and growth. Nucleation could be indicated by a rise in the total number of particles and a decrease in particle size. Crystal growth could be indicated by an increase in particle size with a constant particle count. Controlling PSD is important for several areas of developing pharmaceuticals. Small particles with a narrow PSD are needed for medications with low water solubility. To address these needs, in addition to milling, specialized procedures such as impinging jet crystallization are occasionally used. Different crystal size distributions can increase or lessen the difficulty of the formulation process for medications with high water solubility. Therefore, it is crucial to properly manage crystal size during crystallization. When referring to the solid state, polymorphism is the ability of the same chemical substance to exist in various crystalline formations. Various structures are typically referred to as polymorphs, polymorphic modifications, crystal forms, or forms. Since they relate to separate chemical entities, solvates and hydrates (specifically water solvate) aren't considered polymorphs by this definition. Pseudopolymorphs are another name for solvates and hydrates. Enantiotropy and monotropy are the two types of polymorphism from a thermodynamic perspective. There is no true transition point below the melting point region because, for monotropic forms, one polymorph is the stable solid modification for the whole temperature range up to the melting point. When monotropic polymorphs are taken into account from a thermodynamic perspective, a virtual or projected transition point is produced above the melting point. Given that both polymorphs have a distinct range of temperature below their respective melting points (MPs) when they are in the thermodynamically stable solid phase, enantiotropic crystal forms have a reversible transition point at some temperature.

CHAPTER

5 Analysis of Volatile Organic Compounds

Contents

5.1. INTRODUCTION

Volatile organic compounds or VOCs can be found in natural and artificial or anthropogenic substances, including water, gasoline, and flowers. These substances have a plethora of uses, including those of fuels, solvents, perfumes, biocides, and flavoring agents (Dewulf, Van Langenhove, & Wittmann, 2002). VOCs can also take the form of solids, such as naphthalene and para-dichlorobenzene, which are employed as bathroom deodorizers and vaporizers, respectively. The primary exposure pathway for the majority of VOCs is air (Figure 5.1).

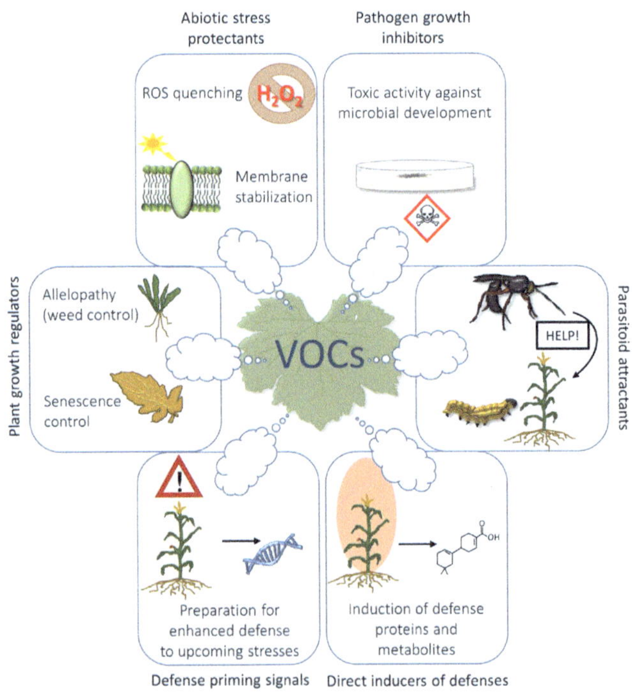

Figure 5.1. Volatile organic compounds.

Source: https://www.frontiersin.org/files/Articles/436279/fpls-10-00264-HTML-r1/image_m/fpls-10-00264-g001.jpg, Creative Commons License.

Food and water are pathways for some naturally occurring VOCs, such as limonene, which is both a natural component of citrus fruits and a synthetic additive to soft drinks, teas, and other products. It has been proposed that soil movement may be important in the case of buildings such as homes or schools constructed in polluted landfills. Acute irritative effects and cancer are the two main health impacts of interest. The latter could have an even bigger economic impact than the first. Pumping a sample of air across a sorbent (activated charcoal) to concentrate the VOCs has been the preferred method for measuring occupational exposures (1–100 parts per million, or ppm, for a given VOC). After that, they are retrieved using a solvent like carbon disulfide.

Active charcoal-based passive badges were made for occupational sampling in the early 1980s (Bertsch & Ren, 2000). The badges, which work on the diffusion principle, are frequently used throughout an 8-hour workday to provide an integrated average exposure for comparison to the occupational norms like threshold limit value, or TLV. But because of VOC residues on the activated carbon left over from the production process, these badges cannot be used for short-term monitoring at environmental concentrations, which are typically measured in parts per billion (ppb). Although various investigations of indoor air pollution have utilized this technique, the high background contamination on the badges can be reduced by extending the sampling period to a week or more. Tenax is a suitable sorbent due to the issues with activated charcoal and difficulties in obtaining accurate recoveries of sorbed compounds. In the 1970s, Tenax, a polymer, was widely used as a more dependable sorbent for ppb levels than charcoal. Tenax has extremely minute background contamination when properly cleaned. Because it can withstand temperatures of up to 250°C, thermal desorption can be used in place of solvent desorption. The concentration is attainable by the sorbent is canceled by solvent desorption because it requires dilution of the VOCs. The inability to retain some volatile organic substances, such as vinyl chloride and methylene chloride, and the production of chemicals, including benzaldehyde and phenol, are drawbacks. Although active (pumped) samplers have been the most common method of using Tenax sorbent, a passive badge incorporating Tenax has also been produced.

5.2. SORBENT

The fine qualities of charcoal and Tenax were attempted to be combined into a multi sorbent system in the late 1980s. Spherocarb and Carbo sieve are two varieties of activated charcoal that were created to offer more dependable recoveries (Ligor, Górecka, & Buszewski, 1998). Activated charcoal was used as a backup sorbent in systems that used Tenax as the primary sorbent. The majority of the VOCs were captured by the Tenax, while the more volatile VOCs were captured by the activated charcoal. Three sorbents, such as Tenax, Ambersorb, and Spherocarb or Carbosieve, were also used in the development of systems. The collection of a wider variety of chemicals and volatilities is possible with all such systems. Whole-air sampling was a technique, which was initially created in the 1970s for upper atmospheric sampling, and does not involve the sorption-desorption process, which should, in theory, reduce the possibility of contamination. However, it necessitates extremely high sensitivity from the detection instruments. Real-time sampling in mobile laboratories with direct air sample injection into a cold trap connected to a gas chromatogram is one possible application of the technique. Another is sampling in evacuated electropolished aluminum canisters for subsequent laboratory analysis. No one technique has emerged as the standard or go-to approach for monitoring VOCs indoors or in the atmosphere. Tenax and evacuated canisters are the two methods commonly used in the US.

The two most popular methods in Europe are Tenax and activated charcoal. Studies that used both techniques discovered consistently greater amounts of total VOCs on the charcoal sorbent. The difference could be caused by extremely volatile organics like pentane and isopentane, which are collected by charcoal yet easily penetrate Tenax. The sorbent methods are well suited for personal monitoring; to obtain a time-integrated sample, a tiny battery-operated pump is used for 8 or 12 hours. However, the whole-air techniques currently in use involve bags or canisters that are too big or heavy to serve as personal monitors. Typically, gas chromatography (GC) is used to separate the components of samples before analysis. Flame ionization detection (FID), electron capture detection (ECD), and mass spectrometry are three commonly used detection techniques (MS) (Gelsomino, Petrovičová, Tiburtini, Magnani, & Felici, 1997). Many compounds can only be accurately identified by GC-MS. Chemicals that coelute or emerge from the chromatographic column at the same time cannot be separated using GC-FID or GC-ECD. Additionally, chlorine and other halogens suppress the GC-FID reaction, making them unsuitable for samples containing halogens. Mass spectrometry can frequently distinguish between compounds that coelute by fragmenting substances and then identifying those fragments. Even GC-MS identifications are frequently incorrect because substances are recognized by comparing these mass fragment spectra to already-existing libraries, and the libraries are imperfect.

5.3. MEASUREMENT METHODS

An alternative to chemical measurement methods has emerged in recent years: the use a trained panel to assess the potential health or comfort effects of an air sample directly. This is due to the complexity of most indoor and outdoor air samples, the cost and inaccuracy of measurement methods, and the complete lack of knowledge of the relationship between measured VOC levels within complex mixtures and resulting health effects. Ole Fanger of Denmark invented this technique, which makes use of panels of 6 to 10 people who have been previously trained by inhaling out unidentified mixes of odorous substances. The judges instantly assess the pollution potential of a test atmosphere when exposed to it, measuring it in decipols (Zweers, Skov, Valbjørn, & Mølhave, 1990). One individual in a room with one air change per hour of ventilation produces one decipol of pollution. The technique can forecast how people will respond to air of a certain quality and estimate the relative contributions of different sources such as the ventilation system, office equipment, and people to indoor air quality. The purge and trap method created by Bellar and Lichtenburg has been the method of choice for sampling VOCs in water. The VOCs are extracted from the water sample using this approach and put into a collection vessel where they are concentrated on a sorbent like Tenax and then evaluated using GC-ECD, GC-FID, or GC-MS. The trihalomethanes (chloroform) are the most critical VOCs to sample in drinking water samples taken from distribution systems that use chlorination as a treatment technique. Detection will frequently be done using ECD, a sensitive technology for halogens. A technique called isotope dilution can be used to evaluate blood.

5.4. EMISSION

Researchers used measurements taken inside buildings or residences to estimate emission rates. They estimated emissions from a variety of personnel by regressing exposure or breath level measurements against the other activities. They extended chamber experiments with paint removers and discovered comparable quantities of methylene chloride. About 5,000 materials used in space missions have had their organic emissions assessed by the US National Aeronautics and Space Agency (NASA) (Wang et al., 2020). Around 3,000 of these materials are used in the regular market. Toluene, methyl ethyl ketone, and xylenes were the compounds released. Aged construction materials subjected to 41-day compartment research revealed that the materials had absorbed roughly 30 VOCs and re-emitted to the chamber over the first 30 days of the study. Even if the source material is eliminated, it can take weeks before sinks in the building stop emitting organic waste again. Another investigation into the sorption of VOCs and SVOCs on materials including plywood and textiles concluded that the sorption was minimal for the VOCs under investigation. Most of the materials emit chemicals at their highest rates when they are first formed. The majority of the total volatile mass for wet materials like paints and adhesives may be released in the first few hours or days after application. According to EPA tests of new buildings, eight of 32 target chemicals assessed within days of the building's completion were 100-fold increased relative to outside levels (Damalas & Eleftherohorinos, 2011). These compounds include xylenes, ethylbenzene, ethyl toluene, trimethylbenzenes, etc. These compounds' half-lives ranged from two to six weeks; thus, it stands to reason that other non-target molecules, like toluene, would have behaved similarly. Paints and adhesives were probably the main contributors. Thus, it would be anticipated that it would take new buildings 6 to 12 months to reach older structures' VOC levels. Emissions from dry building materials like carpets and pressed wood products are likely to stay low for longer periods. With a half-life of several years, formaldehyde from pressed wood products may be slowly released. The primary VOC released from carpets during the first few days is 4-phenyl cyclohexane (4-PC), a reaction product formed in the styrene-butadiene material of carpets. The smell of the new carpet is probably primarily caused by 4-PC.

5.5. APPLICATIONS

Room air fresheners and bathroom deodorants provide a source of carcinogenic and toxic VOC exposure for humans. Since the purpose of these devices is to maintain a high indoor air concentration in the office or home for weeks, prolonged exposures to the VOCs are frequently the ones that most people are most likely to experience. These items mostly contain p-dichlorobenzene, limonene, and pinene, three VOCs that are commonly found in public bathrooms. Many cleaning and polishing solutions also contain limonene and pinene (pine scent), which could result in short-term peak exposures while being used but may not produce as much overall exposure as the air freshener. California's law requires manufacturers of consumer goods to disclose any carcinogenic substances. In 1989 and 1990, indoor air-related bills were introduced in the Senate and the House. Since the EPA classified environmental tobacco smoke (ETS) as a known human carcinogen in 1991, smoking has been banned in many public and private workplaces (Zucco & Doty, 2022). Even in heavily industrialized locations with large concentrations of cars, outdoor air levels of many of the most common VOCs are typically lower than the interior. This fact is not yet fully acknowledged or taken into account by laws. For instance, the US Clean Air Act was renewed in 1990 and continues to primarily address outside air, despite the addition of 189 hazardous chemicals to the list of substances that must be regulated. Numerous of these substances, such as common solvents and household pesticides, are far more abundant and concentrated indoors than they are outside. The primary source of outdoor air levels for many hydrocarbons is gasoline vapor or motor exhaust. For instance, only around 15% of outdoor benzene levels in the US come from fixed sources, whereas roughly 85% of them come from mobile sources. Exposure to certain aromatics like benzene, toluene, xylenes, and ethylbenzene inside of a car can be around six times higher than ambient levels.

5.6. WATER TREATMENT

In regions where drinking water is chlorinated, the water is contaminated by chlorination by-products including chloroform and other trihalomethanes (THMs). The Safe Drinking Water Act or SDWA of 1974 established a maximum of 100 microgram/L for total THMs in water supplies in response to the detection of chloroform in the blood of New Orleans citizens (Rodriguez, 2016). Since the SDWA's passing, several national surveys on THM levels in water supplies have been conducted in the US. About 20% of the 80 treatment plants in the National Organics Reconnaissance Survey surpassed the THM limit of 100 microgram/L. It has become clearer over the years that THM exposure can happen in ways other than through drinking chlorinated water. Showering, bathing, and washing dishes and clothes use treated water, therefore volatilization from these activities is a significant source of exposure. The exposure to chloroform from volatilization during a typical 8-minute shower could be between 0.1 and 6 times higher than that from drinking water from the same source, depending on how much water is consumed daily. These findings are supported by several investigations of full-scale showers, with exposure estimates corresponding to ingesting 1.3–3.7 L/day of tap water at the same concentrations as the shower water.

It has become clearer over the years that THM exposure can occur in ways other than through drinking chlorinated water. Showering, bathing, and washing dishes and clothes use treated water, therefore volatilization from these activities is an important source of exposure. The exposure to chloroform from volatilization during a typical 8-minute shower could be between 0.1 and 6 times higher than that from drinking water from the same source, depending on how much water is consumed daily. These findings are supported by several investigations of full-scale showers, with exposure estimates corresponding to ingesting 1.3–3.7 L/day of tap water at the same concentrations as the shower water. It was also discovered that breathing in airborne chloroform was equal to ingesting it. Chloroform and other THMs have been identified in ice cream and soft drinks that are manufactured with treated water. In animal studies, some naturally occurring elements of fruits and vegetables, including limonene, have been proven to be mutagenic and/or carcinogenic. Tetrachloroethylene from dry cleaners and trichloroethylene from the glue in the containers have both been discovered to be absorbed by butter and margarine in supermarkets. A direct indication of the overall dosage through all environmental pathways is given by measuring body burden. When the measurement of a single body fluid is combined with adequate pharmacokinetic models, including knowledge of how the VOC partitions between blood, air, and fat, it can indicate the total body burden and occasionally even the mode of exposure. Measuring exhaled breath is the accurate way to determine the body load of various VOCs. When employing evacuated canister samples or Tenax sorbents, detection limits are often much below 1 microgram/m^3. Blood tests would have to be sensitive enough to identify concentrations as low as 10 ng/l to investigate. These measurements show how important smoking was as a source of benzene and styrene exposure. While it is true that breath measurements only capture the most recent exposure, they can also capture long-term exposure provided the timing of the measurement is properly considered. Some experimental studies have established the residence periods of more than a dozen VOCs in various body compartments, including blood (2–10 min), tissues rich in blood vessels (1–2 h), and other tissues (4–8 h). Blood tests have been performed at occupational levels for a long time. Recent studies have measured blood levels of benzene and chloroform at environmental levels.

5.7. VOC

VOCs can aggravate the eyes, nose, and throat, and cause reactions in the central nervous system like headaches, dizziness, and short-term memory loss. These signs are present in newly constructed or undergoing refurbishment buildings. VOCs are suspected, even though sick building syndrome's (SBS) causes or reasons are still unknown. Indoor VOC levels are frequently high. Levels are even greater in newly constructed or refurbished structures (Du,

Mo, Zhang, & Xu, 2014). According to field research, sick buildings are frequently found to have total VOC levels surpassing 1–2 mg/m^3. According to laboratory research, sensitive and healthy persons can both experience some symptoms like headache and eye irritation at TVOC levels of 5 and 25 mg/m^3. Others cause cancer in animals and may do the same to humans (methylene chloride, trichloroethylene, tetrachloroethylene, chloroform, dichlorobenzene). Some are weak animal carcinogens or mutagens. One way is to extrapolate from animals to people and from high doses to low doses, assuming that there is no threshold, to determine the carcinogenic potency of a particular chemical. An effective threshold might be reached, for instance, if one species has an entirely different metabolic pathway from another or if it has a stronger mechanism for DNA repair. Due to these ambiguities, a specific animal carcinogen may not at all induce human cancer; in fact, the cancer risk may be zero.

The use of short-term animal toxicity studies to calculate carcinogenic potential is an important method. This is predicated on the concept that toxicity and carcinogenicity are connected. Given that more toxicity research has been conducted than carcinogenicity studies, if the assumption is correct, this significantly expands the pool of chemicals that can be taken into account. The same concerns still exist, in addition to the uncertainty surrounding the fundamental presumption that toxicity is connected with carcinogenicity. The Environmental Protection Agency or EPA in the US has systematically used the techniques to estimate upper limits of the carcinogenic potencies of more than 50 compounds, including several VOCs. The majority of the total airborne hazard associated with most of these compounds was caused by indoor sources. Due to its exclusion from consumer items, carbon tetrachloride is the only one of the target compounds for which outdoor sources account for the majority of the airborne risk. Only benzene is thought to be a human carcinogen among these substances. As a result, benzene's risk assessment is based on more reliable data than any other. The only chemical with human epidemiological studies indicating a potential link between environmental exposure levels and cancer risk is benzene. Because there are so many VOCs, it is hard to discuss every chemical and its toxicity. The goal is to include the most significant representatives of the various classes of VOCs in terms of both their incidence and their dangerous qualities. The toxicokinetic, biotransformation, and toxicity following oral intake and exposure by inhalation, particularly in experimental animals, will be important to discuss. The most significant method of exposure for VOCs appears to be inhalation, however oral and cutaneous exposure is also significant for some VOCs. In many instances, only one exposure route, typically oral exposure but occasionally includes inhalation can be accurately represented by toxicological data. As a result, understanding the toxicokinetic behavior of some VOCs following ingestion, absorption, or cutaneous exposure helps to better understand the toxicological profile of related VOCs. It is crucial to determine whether a VOC's absorption which may vary for exposure routes leads to comparable distribution of the VOC and its metabolites throughout the bodies of animals and people, and which physical and chemical characteristics are crucial to the distribution. Among these characteristics, lipophilicity and water solubility are crucial for a compound's behavior inside the body. Biotransformation is frequently crucial for the detoxification or activation (toxification) of VOCs, in addition to dispersion. In many instances, the liver is particularly vital in the modification of VOCs such that the metabolites are directly water-soluble or capable of conjugation with other substances made by the body (Li, Pal, & Kannan, 2021). To make the metabolites more soluble, the metabolites might also go through reduction. Phase I reactions include detoxification/toxification processes, which frequently involve the enzymes found in liver microsomes. Phase II reactions include conjugations to, for instance, glucuronic acid. Once a VOC's metabolites are water soluble, they can be expelled through the urine excretion pathway or breathing. The entero-hepatic route may be used to remove the lipophilic substances and metabolites. Kidney function has a big impact on renal elimination. Elimination may be hampered if this function is compromised by VOC or its metabolites. For elimination by exhalation, lung function and gaseous exchange in the lungs are crucial. The removal of a VOC by exhalation is affected and the toxicokinetics of the VOC at toxic exposure levels are complicated if a VOC has a direct toxic effect on the lungs.

5.8. TOXICITY

The toxicokinetic behavior in terms of both quantity and quality substantially influences the occurrence and onset of the toxic characteristics of VOCs. Varying exposure routes may result in distinct concentrations of a VOC or its metabolites, and thus, different harmful effects on the target organs. Information on internal exposure is provided by toxicokinetic. This internal exposure threshold can be used in physiologically based models for risk evaluations when there is a direct impact on the relevant organ or tissue, provided that data on the mechanism of the VOC's toxicity in the tissue is available. However, compared to other effects of the substance, some hazardous consequences are less significant. Smoking, physical activity, stress, and other variables, such as diseases affecting the respiratory system, can alter these conditions. The degree of lung ventilation and blood flow in healthy individuals is mostly controlled by metabolic rate. The amount of physical activity has a role in the metabolic rate in typical conditions. When compared to uptake during exercise, a gas's lung uptake at rest is very different. It was demonstrated that the concentration in alveolar air increased when alveolar ventilation increased in physical activity. The arterial blood levels of VOCs were greater during activity than at rest. Blood solubility and the rate of increase for the VOCs measured were related. While the blood/air partition coefficients for some VOCs (methylchloroform, methyl chloride, trichloroethylene, and toluene) were not altered during exercise, they significantly increased for the aromatic components of spirit. Adipose tissue has minimal blood flow while at rest, and this blood flow is likely even lower when exercising. Exercise may cause a decrease in blood flow to the liver, the gastrointestinal tract, and the kidneys. However, coronary blood flow also rises along with blood flow to the lungs, neurological system, and brain. Therefore, physical activity creates conditions that are favorable for uptake in the lungs, minimal delivery to some organs, and massive delivery to others. The relatively high blood concentrations of VOCs discovered were a result of this effect of physical activity. While all VOCs examined showed a drop in arteriovenous differences during vigorous activity, the overall amount released from arterial blood per unit of time was higher. Humans have demonstrated a rise in pulmonary absorption during exercise in previous methylene chloride investigations. Exercise increased the overall amount of toluene that people inhaled.

5.9. TRANSFORMATION

The biotransformation of VOCs, which mostly occurs in the liver and lungs after exposure by inhalation, may be influenced by variables affecting the organ's metabolic capabilities. These variables include induction time, substances in diet or the environment, exposure dose, and exposure duration. Researchers demonstrated using inhalation tests that pyrazole pretreatment effectively reduced *in vivo* styrene metabolism, whereas phenobarbital pretreatment significantly enhanced the maximum metabolic velocity of metabolism. This demonstrates that the same variables that prevent the biotransformation of xenobiotics consumed orally also affect

VOCs once they are absorbed. The observed effects on toxicokinetic and biotransformation highlight the significance of these influencing elements (stress and physical activity) in inhalation investigations for toxicological research. Exhalation also contributes to the removal of a VOC and its metabolites if they are volatile (Weisel & Jo, 1996). Therefore, the same factors that affect a VOC's intake may also affect how much of the VOC and its metabolites are excreted. Once VOCs are converted to water-soluble metabolites, the key method of removal is by urine excretion; therefore, disorders affecting kidney function may prevent the metabolites from being excreted. The entero-hepatic route may be used to eliminate lipophilic VOCs or metabolites, however, there are differences. Therefore, variables affecting bile acid synthesis may control this mode of elimination.

Water solubility, fat solubility, the VOC's carrier, the content of the meal, and the composition of the gut bacteria are factors that affect uptake when it comes to oral exposure to VOCs. The route of movement through the body has an impact on how a VOC and its metabolites are distributed. This may primarily be the lymphatic system in the case of lipophilic chemicals, as the VOC doesn't reach the liver for biotransformation until it has circulated in the blood. The respiratory system is where the VOC first enters, where, depending on its volatility, it may be partially exhaled. In contrast, water-soluble VOCs are directly subjected to biotransformation activities when they reach the bloodstream through the liver's renal portal system. Most biotransformation will occur in the liver. Therefore, factors affecting the liver's metabolism will either improve or worsen VOC removal by biotransformation. Although there may be differences in the amounts lost by exhalation, urine, and feces, the removal of VOCs and their metabolites after oral consumption will follow the same routes of elimination as after exposure. Water solubility, fat solubility, dehydrating properties, the vector in which the VOC is dissolved, the skin condition, and occluded or non-occluded exposure to the VOC are factors that affect the uptake of VOCs through the skin in the case of dermal exposure. Toluene biotransformation in humans and animals is expected to be comparable. The microsomal oxidase system appears to convert toluene's side-chain to benzyl alcohol as the first step in the metabolism of toluene to benzoic acid. In contrast to non-induced rats, the induction of hepatic oxidase by phenobarbital accelerated the metabolism of toluene and led to high levels of benzoic acid in the blood. Toluene biotransformation appears to take place largely in the liver.

5.10. METABOLISM

The development of metabolites in the liver, kidneys, and blood of mice within minutes of exposure to toluene and increased urine excretion of hippuric acid in rabbits within 0.5 hours of inhalation demonstrate the quick metabolism of toluene. Benzoyl glucuronide can be created by conjugating benzoic acid with glucuronide acid. O-cresol and p-cresol are produced as a result of ring hydroxylation by oxidases, albeit this process only takes place in very small amounts. In humans, 10–20% of the absorbed toluene is excreted as the glucuronide conjugate, and at most 1% is excreted as 0- or p-cresol, either as glucuronide or as sulfates. The kidneys excrete 60–70% of the absorbed toluene as hippuric acid. After exposure to toluene at values between 187 and 375 mg/m^3 for many hours, nine of its metabolites reverted to background levels in less than 24 hours. Normal sources of phenol and conjugated phenols include human and animal tissues, urine, feces, saliva, and perspiration. The food composition affects the amount of phenol in the blood. Regardless of the method of exposure, phenol is readily absorbed. While damage to the skin depends on the concentration of the phenol solution, the route of absorption through the skin is mostly dependent on the exposed surface. Humans mostly absorb phenol vapors and solutions through their skin.

The liver has the highest concentration of phenol 15 minutes after receiving 0.5 g/kg of body weight, followed by the blood, central nervous system, lungs, and central nervous system (Afonso et al., 2013). The phenol has been evenly dispersed throughout all tissues after 82 minutes. With time, the ratio of conjugated phenol to phenol steadily rises. Following oral administration of 14C-phenol to rats, the liver was found to have the highest value relative to plasma (42%) followed by the spleen, kidney, adrenals, thyroid, and lungs, with peak values occurring after 30 minutes. The highest amounts in tissues were seen in the kidneys and liver two hours after intravenous treatment in mice and rats. Phenol is broken down in the liver into carbon dioxide (23%), minor amounts of catechol, and hydroquinone. It is then expelled in the urine (72%), both unaltered and as conjugated phenols, excreted in the feces (1%), and exhaled in very little amounts as phenol. Just 4% is estimated to be retained in the body. Numerous phenolic metabolites, including phenyl sulfate, phenyl glucuronide, chino sulfate, and chino glucuronide, have been discovered in the urine of various animal species. Pigs and cats showed a restriction in the conjugation of phenol with sulfate or glucuronic acid. The level, route, and species of the animal affect the excretion rate. The rate of excretion is fastest in rats, where 95% of phenol provided orally at a dose of 25 mg/kg body weight was eliminated by urine in less than 24 hours. T

he rhesus monkey experienced the slowest excretion (91%). Phenol or its conjugates are eliminated through the bile, perhaps as a result of restricted urine excretion. Pentachlorophenol (PCP) is the frequently mentioned chlorophenol in data on its kinetics and biotransformation. PCP's volatility is constrained, though, and issues with inhalation are mostly limited to aerosols. It is anticipated that all chlorophenols are rapidly absorbed and excreted, with urine being the main mode of elimination, based on the information at hand and the physicochemical characteristics of chlorophenols. After cutaneous or inhalatory exposure, there are no specific absorption studies with mono and dichlorophenols (DCPs) available. Chlorophenols are, however, easily absorbed via cutaneous and inhalatory exposure routes in experimental animals and humans.

The liver had the highest quantities of 2,4-dichlorophenol (2,4-DCP) or 2,4,6-trichlorophenol (2,4,6-TCP) following a single parenteral injection. The liver and kidneys of cattle and sheep given 2,4-dichlorophenoxyacetic acid for 28 days had the highest and second-highest 2,4-DCP contents, respectively. The lower chlorinated chlorophenols (monochlorophenol (MCP), dichlorophenol (DCP), and trichlorophenol (TCP)) are present in tissues and body fluids primarily as glucuronide and sulfate conjugates, according to studies on biotransformation using different laboratory animals, different routes of exposure, and different compounds. There is limited accessible quantitative data on the excretion of MCP, DCP, TCP, and their metabolites. When benzene was supplied to rats, dispersion took place within 2–6 hours, with blood, adipose tissue, and bone marrow having the highest quantities.

The liver, spleen, bone marrow, and blood of mice had the highest levels of benzene after 3 hours of cutaneous contact. One hour after cutaneous application, the CO_2 content peaked. The primary method of excretion for benzene in dogs, rabbits, mice, and rats is through exhalation. Up to 50% of benzene is exhaled out of the body, while the remaining portion is converted to phenol by the oxidase system.

Benzene-epoxide, hydroquinone, minute amounts of the dihydroxy derivative hydroxyquinone, and catechol are among the intermediate metabolites of benzene (Sorokin, 2013). The principal forms of excretion for these hydroxy derivatives into urine are glucuronide or sulfate esters. The proportional amounts vary depending on the animal species. The main metabolites in rabbits and mice are glucuronides, but the main metabolites in rats are sulfate esters. The entire body is exposed to chloroform. The concentrations in fat, blood, liver, lungs, kidneys, and the nervous system are at their highest. The liver and kidneys (particularly the renal cortex of male mice) exhibit tissue component binding. It is believed that a cytochrome P_{450}-dependent enzyme system converts chloroform into phosgene, which primarily combines with cysteine or glutathione. The tissue macromolecules may also bind phosphogene. It has not, however, been conclusively demonstrated that chloroform may produce phosgene. The production of carbon monoxide, which is cytochrome P_{450} and glutathione-dependent, is a minor biotransformation process of chloroform. When exposed at low levels, phosgene appears to be the primary toxin, however when exposed at high levels, chloroform itself appears to be a contributor to the toxic effects (anesthesia). Chloroform does not appear to produce trihalomethyl radicals, in contrast to other halomethanes. The liver is the main organ for metabolism. Chloroform is metabolized by humans and monkeys to a lesser amount than it is by mice and rats.

There is a sex variation in metabolic activity; male mice exhibit significant renal metabolizing activity. The lungs are responsible for eliminating carbon dioxide, the primary metabolite of chloroform. Excretion through other means is hardly noticeable. Chloroform seems to be eliminated very slowly through the lungs, excreting just a small percentage of the body. Retention is greater in overweight people. The amount of chloroform breathed is a good indicator of metabolic rate. Chloroform is expelled at relatively substantial levels by slow metabolizers.

It has been reported that the terminal plasma half-life of chloroform in humans is about 90 minutes. The $t_{1/2}$ in exhaled air, however, may last for up to two days. These results showed the presence of a deep fatty compartment where chloroform is slowly eliminated. Due to hepatic pulmonary effects, only 55% of an oral dosage is accessible in humans. Rats were inhaled with 14C-methylene chloride for 6 hours, resulting in tissue concentrations that ranged from 7 to 23% of the body. After 2 hours of exposure, blood levels of the radioactive label achieved a stable state and started to fall dramatically. The liver, kidneys, and lungs had the highest amounts. High amounts of methylene chloride were found in the liver, kidneys, white adipose tissue, adrenal glands, and blood one hour after inhaling a high concentration of the chemical. Within 2 hours of exposure, the elimination of 25–90% of the accumulated radioactivity from these tissues was seen. Methylene chloride was found in brain tissue and perirenal fat in conditions of extended exposure (Rioux & Myers, 1988). In mice, the distribution was comparable. According to data from distribution studies using radioactively labeled methylene chloride in rats and mice, the liver and kidneys accumulate radioactivity more than other tissues whereas the brain and body fat eliminate it more quickly. Humans exposed to methylene chloride through inhalation experienced a slow rise in blood concentration. The adipose tissues also showed an increase in methylene chloride.

5.11. ENZYMES

Enzymatically catalyzed, the biotransformation of methylene chloride (Figure 5.2) exhibits rate-limited features at high concentrations. At low concentrations, first-order kinetics is shown, but at large doses, zero-order kinetics is seen. The P_{450} system, which produces carbon monoxide, seems to be a high-affinity, low-capacity pathway. At all available exposure concentrations, the glutathione-mediated route (glutathione transferase) is a low-affinity, high-capacity first-order activity. The combined effect of routes results in maximal metabolic velocity. The difference between the supplied dose and the effective dose increases with exposure because the metabolism of methylene chloride is saturable.

Although the main method of elimination for inhaled methylene chloride in rats and mice was urine excretion of non-volatile metabolites, a small amount was also found in feces. Less than 5% of the 70% of methylene chloride absorbed during inhalation exposure was exhaled, showing that biotransformation plays a significant role in removal.

The metabolite excretion levels in the urine were related to the amounts of inhalational exposure. After oral and inhalation exposure, monochlorobenzene (MCB) is well absorbed. After cutaneous exposure, there appears to be little absorption. All tissues analyzed contained MCB shortly after oral treatment, with adipose tissue possessing the highest concentrations. Four-chlorophenol, four-chlorocatechol, and four-chlorophenylmercapturic acid were the main metabolites of MCB. Only trace levels of metabolites were found in the feces, with the majority being eliminated in the urine. Ethereal sulfates (33.9%), glucuronides (33.6%), mercapturic acids (23.6%), diphenols (4.2%), monophenols (2.8%), and 3,4-dihydro-3,4-dihydroxychlorobenzene (0.6%) were the urinary metabolites of MCB (Kusters & Lauwerys, 1990).

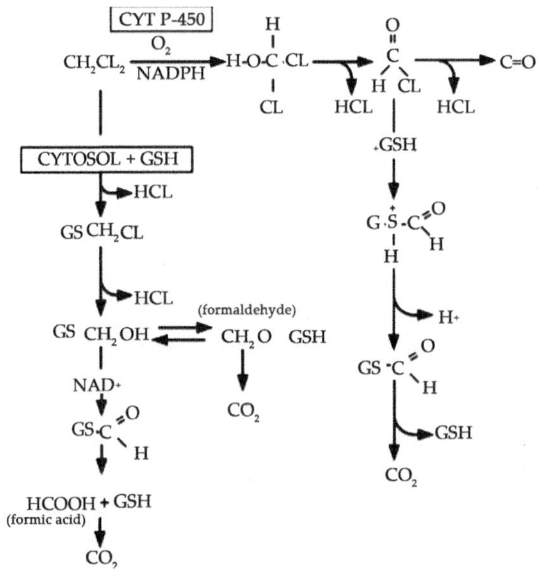

Figure 5.2. Biotransformation of methylene chloride.

Source: https://u.osu.edu/murray.1179/files/2020/07/Screen-Shot-2020-07-07-at-8.03.59-PM.png, Creative Commons License.

5.12. EXPOSURE

Dichlorobenzenes (DCBs) are absorbed by several modes of exposure. Tissue distribution after oral inhalatory and subcutaneous exposure appeared to be comparable (Monferran, Pesce, Cazenave, & Wunderlin, 2008). The adipose tissue, liver, kidneys, and lungs had the highest levels. Male and female rats were shown to have different organ distributions following inhalation exposure to 1,4-DCB; female rats had significantly higher amounts of 1,4-DCB in their livers, but male rats' kidneys had significantly higher levels compare to the female rats. In order to produce 3,4- and 2,5-dichlorophenol, respectively, the 1,2-DCB and 1,4-DCB were primarily oxidized. Around 50% to 90% of the injected dose was eliminated in the feces as metabolites within 24 hours of exposure. Glucuronides, ethereal sulfate (21%), mercapturic acids (5%), 3,4-dichlorophenol (30%), and 2,3-dichlorophenol were the primary 1,2-DCB excretion products. The excretion of DCBs in the feces was minute. Through the entero-hepatic circulation, the DCBs were absorbed. After exposure by ingestion, inhalation, and skin contact, trichlorobenzene (TCBs) are absorbed. After oral administration, the liver, kidneys, adipose tissues, bladder, and gastrointestinal tract of rats contained the highest concentrations. Skin and muscular tissues were also said to have higher contents. The levels of 1,3,5-TCB in tissues were highest after exposure. TCPs, which are produced by the breakdown of TCBs, are then conjugated with glutathione in rats. In contrast, although the intermediate metabolite arene oxides is hydrolyzed, monkeys do not utilize glutathione in the metabolism of TCBs. Dihydrodiol produced as a result is eliminated as a glucuronide. Rats expelled almost 70% of the substance they had consumed orally in their urine and 15% of it in their feces after 24 hours, compared to 40% in the urine and less than 1% in the feces of monkeys. The entero-hepatic cycles were also used to reabsorb TCBs.

VOCs generally have low to moderate acute oral toxicity. The oral LD_{50} is 200 mg/kg body weight (bw) or greater (lethal dose causes 50% mortality) (Korpi, Järnberg, & Pasanen, 2009). Sometimes toxicity increases following cutaneous or inhalatory exposure. Depending on the exposure route, toxic effects at lethal dose levels can vary. The distribution and metabolism of VOCs may help to explain this variation in toxicity. The lungs are implicated in situations involving inhalatory exposure. At this lethal dose levels, hepatotoxicity and renal toxicity are frequently involved, despite the fact that different VOCs have varied toxicological profiles. While certain VOCs have impacts on the central nervous system, others have anesthetic effects. According to studies, the oral LD_{50} for benzene in rats ranges from 3,900 to 5,970 mg/kg bw and the inhalatory LC_{50} is between 15,000 and 60,000 mg/m^3. Although there is no dermal LD_{50} available, leukocytopenia following subcutaneous exposure has been demonstrated in mice and rabbits at doses of at least 5,000 mg/kg bw. When exposed orally, stomach lesions were also seen. Styrene's oral LD_{50} in rats ranges from 2,150 to 8,057 mg/kg body weight. This is more than the 316 to 660 mg/kg range for mice. Styrene has been shown to cause liver and kidney damage at lethal dose levels, as well as pneumonia and effects on the central nervous system when inhaled.

The oral LD_{50} for toluene in rats ranges from 5,000 to 7,500 mg/kg bw. When exposed for 4 to 7 hours, mice and rats' inhalatory LD_{50} for toluene ranges from 26,000 to 72,000 mg/m^3 (Withey & Hall, 1975). The range of the lowest recorded lethal dose after toluene exposure after subcutaneous injection was 1,100–8,700 mg/kg bw. At these lethal dose levels, there were toxic consequences that included tremors, increased neurotransmitter levels, and loss of coordination. For combined xylenes, the oral LD_{50} in rats ranged from 4,300 to 8,000 mg/kg bw. The acute toxicities of the various xylene isomers change when exposure occurs through breathing. The LD_{50} for a-xylene, m-xylene, and p-xylene in rats was 4,595 mg/m^3, 5,267 mg/

m^3, and 3,907 mg/m^3, respectively. The range of the LD_{50} for xylene and ethylbenzene combinations was 6,350 to 6,700 mg/m^3. Cats exposed to 9,430 mg xylene/m^3 showed characteristic signs of the central nervous system, including salivation, ataxia, tonic, spasms, etc. For m- and p-xylene, the fatal subcutaneous dose in rats varied from 4,000 to 8,000 mg/kg bw, while for a-xylene, it was between 2,000 and 4,000 mg/kg bw. While the inhalatory LD_{50} for phenol in rats was > 900 mg/m^3, the oral LD_{50} for phenol in mice, rats, rabbits, and dogs ranged from 300 to 600 mg/kg bw (8 h exposure). The dermal LD_{50} values for rats and rabbits were 670 and 850 mg/kg bw, respectively. Tremors, convulsions, and an elevated heart rate were the toxic effects at these lethal dose levels, which were followed by a low and irregular pulse, a rise in salivation, and a drop in body temperature. The throat and esophagus tissues displayed erythema, and necrosis after oral absorption. Lung lesions, infarcts, bronchopneumonia, bronchitis, and hyperplasia of peri bronchial tissues were observed after inhalation exposure.

Other consequences at this high phenol level included demyelination of nerve tissues, myocardial necrosis and degeneration, liver, and kidney damage. The majority of symptoms at sublethal exposure levels of chlorophenols are comparable and include tremors, dyspnea, convulsions, a rise in body temperature and respiration rate, as well as a variety of motor weakening symptoms. In contrast to the uncoupling of phosphorylation, which was higher for phenols with a higher degree of chlorination, tremor activity was high for phenols with a low degree of chlorination. When chloroform was dissolved in sesame oil, the oral LD_{50} values were 1,120 and 1,400 mg/kg bw, respectively. When chloroform was dissolved in aqueous solution, the oral LD_{50} values were 353 and 1,366 mg/kg bw. Between 10,000 and 100,000 mg/m^3 was the deadly inhalatory concentration in rabbits and guinea pigs. Anesthesia, cardiotoxicity, respiratory acidosis, and hepatotoxicity were the toxic consequences following inhalation exposure at these sub-lethal dose levels. Hepatotoxicity and renal toxicity were the main consequences of oral chloroform intake. Hepatic and renal damage were among the toxic consequences following cutaneous exposure to 75–1,500 mg/kg bw.

Only a few studies with toluene administered orally or topically have been conducted; the majority of short-term toxicity studies with toluene have been conducted to examine the consequences of repeated inhalation exposure. The most popular test animals have been rats and mice. Several studies showed an increase in liver weight but no alterations in liver histopathology. At 9,400 mg/m^3 of inhalation exposure, oxidase enzymes were observed to be induced in the liver. At treatment levels of 2,280 mg/m^3, rat kidney collecting tubules underwent changes. The liver did not show any histological alterations; however, the hepatotoxic enzymes ALAT and ASAT were identified at inhalatory dosage levels between 12,000 and 45,000 mg/m^3 (Woutersen, Appelman, Wilmer, Falke, & Feron, 1987). There were minor alterations in the bone marrow and blood in both mice and rats subjected to 3,800 mg/m^3 of the substance. Most xylene short-term toxicity studies use inhalation tests, where rats exposed at exposure levels of up to 1,150 mg/m^3 showed only minimal impacts on their red and white blood cells. After inhaling 300 mg/m^3 for 64 days, there was minimal liver damage and pulmonary irritation in guinea pigs. In subcutaneous tests, rats subjected to 800–1,600 mg/kg bw/day showed relatively minor decreases in red blood cells. Xylene is not myelotoxic and generally does not appear to have negative effects on hematopoiesis. Although these effects appear to be reversible when exposure is stopped, at high exposure levels certain hematological alterations such as a reduction in red blood cells, leucocytes, lymphocytes, and thrombocytes have been seen.

There is not much proof that xylene harms the liver, kidneys, or lungs, despite considerable animal research. There haven't been many reports of hepatic, renal, or pulmonary impairment. In recent experiments, rats exposed to 860 mg xylene/m^3 for up to 13 weeks did not exhibit any liver, kidney, or lung damage based on blood chemistry or histological analysis. Rats exposed to m-xylene vapors showed enzyme stimulation of the cytochrome P_{450} system and associated mono-oxygenase in the liver and kidneys, but no increase in liver or kidney weight, and no biochemical or histological evidence of hepatotoxicity were noticed. Adaptive rather than harmful, these inductive effects are typical of xenobiotics like alkylbenzene. No alterations in hematological markers or histological changes in the liver or kidneys were seen in a short-term oral toxicity trial in rats with 43, 131, 435, and 1,305 mg xylene/kg bw, however, growth retardation and water retention by the kidneys were detected. Hepatotoxicity and renal toxicity were seen in short-term damage experiments using phenol given orally to rats and mice. From 50 mg phenol/kg body weight onward, the hazardous effects were evident. At 50 mg/kg body weight following subcutaneous injection to rats, effects on leucocytes were also noted. Cats were particularly sensitive to the harmful effects of phenol on the neurological system. Dermal toxicity of phenol in rabbits indicated local skin irritation at dose levels of 100 mg/kg bw and systemic effects like tremors, and death at 130 mg/kg. After inhaling high quantities of phenol (100–200 mg/m^3), short-term toxicity tests on rats, rabbits, and guinea pigs showed effects on the lungs and heart, and to a lesser extent on the liver and kidneys. Acute lobular pneumonia, an arterial vascular system lesion, and lesions of the liver and kidneys were discovered during the histopathological examination. Guinea pigs that were exposed to 3.8–152 mg/m^3 have severely altered respiratory systems. Chlorophenol oral short-term toxicity trials are primarily research intended to establish a safe dosage range for teratology, reproduction, or chronic toxicity studies.

CHAPTER

6 Electrochemistry

Contents

6.1. INTRODUCTION

The use of electrical energy to impact organic reactions has been explored for a very long time and in a variety of ways. When larger electrical equipment was first introduced into scientific and technical industries, various challenges presented by organic chemistry could finally begin to be seriously addressed. For over a decade, organic electrochemistry has been developing quietly but steadily (Little, 2020). Chemical reactions can be carried out either directly or indirectly using electrical energy. Directly using electrical energy requires that the field the current traverses be of an electrolytic nature. Indirectly using electrical energy requires that it be transformed into another form, such as heat or light, which can cause chemical phenomena outside the current field. Theoretically and practically, both usable forms of electricity are significant, the former in electrolysis, particularly in reduction, oxidation, and substitution reactions, and the latter in pyrogenic and photochemical processes. The silent discharge is another type of electrochemical reaction in which the relationship between electrical work and the chemical product is still unclear. Despite the limited information that is now available regarding this type of electrical energy, it can be said with certainty that it plays a crucial role in the synthesis of simple organic bodies and may even be a way to explain the processes that nature uses to create materials. By far, the most extensive research has been done on electrolytic reactions, notably the reduction processes. Successive studies on oxidation and substitution reactions have been less common. Numerous references have been made to the pyrogenic decompositions and syntheses of organic compounds caused by induction sparks, electric arcs, or highly heated conductors. However, in this line of effort, scientific discoveries are still in their infancy. We know very little about how carbon compounds are affected by convective discharge.

6.2. ELECTROCHEMISTRY

This differential treatment and the product are explained by the diverse characteristics of organic entities. Carbon compounds are often reduced at certain groups in the molecule without completely destroying the latter. Usually, the entire molecule is exposed to the electrolytic oxygen activity. When a substance is reduced, the final product is chemically very similar to the initial substance; when a substance is oxidized, the ultimate product is the entire combustion of the molecule. There are a lot of alternatives between an oxygen-only attack and the whole oxidation of a molecule. If at all feasible, this must be realized by the observation of predetermined experimental conditions, which is frequently challenging (Moeller, 2018). As a result, oxidation operations are more complicated and typically less profitable than reduction processes. The same arguments also hold true for electrolytic replacement, an anodic process that is very weakly shielded from the oxidizing effects. Electrothermic decompositions and syntheses of organic bodies were restricted to a small area because carbon compounds are sensitive to high temperatures, provided that the heat source is an electric arc. Since metallic wires or carbon filaments have been employed as heat sources that are easily controlled by changing the pressure, electrical energy has proven to be a useful tool for examining the behavior of sensitive materials at reasonably high temperatures. For organic electrochemistry, specific types of experiments are required due to the characteristics of both electric energy and carbon molecules. These completely diverge from experimentation that is done just in the chemical field, necessitating the use of novel experimental techniques. The perspectives on selecting the ideal settings for an experiment got more apparent as it became possible to identify the critical elements of an electrochemical process.

New experimental settings were produced as a result of theoretical studies. By meeting specific criteria, the theory comes to be related to both the experiment's subject and its design. In addition to the chemical products, a description of organic electrochemistry must take into account theory and methodology. Three processes electrolytic, electrothermic, and electric-discharge reactions can be distinguished based on how electrical energy is used in organic chemistry. In order to differentiate between the two possibilities, biological bodies must be electrolyzed. The ions that regulate the conductivity are provided in the first instance by the chemical itself. The task of electrolysis, therefore, consists of moving these ions to the anode and cathode; whether they are released molecularly or atomically, or whether they interact with other ions, with the constituent still present in the solution, with each other, or with the solvent is not of great concern. Since organic cations, such as the organic ammonium ions, have not been extensively studied with regard to their behavior in electrolysis, the anions are essentially the organic ions that are considered. Since ions cannot exist without their electrical state, the actual release of the ions cannot be seen. Instead, upon their discharge, the anions frequently interact with one another. Thus, either a union of several anions occurs or frequently, these reactions are accompanied by more complex transpositions and decompositions. Under specific conditions, fatty acid anions exhibit this behavior to a greater or lesser extent. But in order to carry out the electrolysis, other ions must be present if the organic component does not transmit the current. Typically, an inorganic acid, basc, or salt is dissolved in the solution for this purpose. Corresponding organic molecules may also be utilized. The organic non-electrolyte is then mostly unaffected by the flow of electricity (Fidaleo & Moresi, 2011).

To the electrodes, where they can discharge themselves, only the ions are transported. It is important to note that the carbon molecule combines with the released ions to transform into a depolarizer. When ions are discharged that readily react with a variety of organic acids, bases, and salts, they can act as depolarizers. For instance, p-nitrobenzoic acid is gradually converted to p-azobenzoic acid in an alkaline solution. Since the sodium ions in the sodium hydroxide solution handle the conductivity, the nitrobenzoic acid essentially behaves as a depolarizer rather than an electrolyte as a result of the sodium ions being discharged. Ions that work on an organic depolarizer can be provided by organic electrolytes. Alcohol serves as both a depolarizer and a solvent in this situation. As a result of the electrolysis, the organic bodies either discharge and eventually undergo an additional reaction with their ions at the electrodes or operate as depolarizers. It is extremely rare for a compound to operate as both a cathodic and anodic depolarizer at the same time, therefore it can be further split into the two classes of cathodic and anodic depolarizers. More frequently, a cathodic (or anodic) depolarizer gains the ability to depolarize anodically by reacting with the cations (or anions). Cathodic reduction can transform a reducible body into one that is easily oxidized, or open to the action of anions.

6.3. DEOXIDATION

Organic cations, hydrogen, and metal ions travel to the cathode in cathodic depolarizers. Deoxidization is a process in which hydrogen and metals remove oxygen from a chemical. Hydrogen can also be given directly to the product. Reducible chemicals are those that can produce oxygen, absorb hydrogen, or do both at once. As a result, they are oxidizers in and of themselves, and one of their primary functions is to eliminate positive discharges. Reduction describes the reaction occurring at the cathode. The electrolysis reduces each cathodic depolarizer. At the anode, the conditions are a little complicated. All anodic depolarizers are indeed oxidizable, including reducing agents that eliminate negative charges (Turrentine, 2002). However, because of the distinct variety of the anions, the reaction is more diversified at the anode than at the cathode. Oxidation is the term used when an action solely entails the removal of hydrogen and the addition of oxygen, or both. It is possible that the anion experiences alterations before it interacts with the depolarizer, in which case the organic component is no longer considered to be a true depolarizer for the anion but just for the by-products of its disintegration. The anion CH_3COO would behave in such a way in the presence of a base that only carbonic acid, when divided into ethane and carbon dioxide, would react with the base. Such a process, however, can no longer be categorized as electrochemical. Since there are so many different ways that an electrolytic reaction could occur, it appears especially challenging to ascertain its nature. In the electrolysis of organic compounds, another type of reaction takes place. The product might first experience an unstable modification before swiftly undergoing additional changes, depending on whether it happened first or second. Intermediate phenylhydroxylamine is formed during the reduction of nitrobenzene in concentrated sulfuric acid because aminophenol is produced readily.

The unstable modification can be separated from the original electrolytic fluid by mixing with benzaldehyde. It takes longer for the aldehyde to react with the intermediate product, phenylhydroxylamine, than it does for sulfuric acid to rearrange molecules. By including various chemicals in the electrolytes that react more quickly with the phase than the oxidation or reduction can take place, intermediate phases of electrical oxidation and reduction can be isolated similarly. The electrolytic approach stands apart from all other chemical processes in several important ways. First, the current creates effects that the chemical approach is unable to create without the aid of specific substances, such as lead peroxide, chromic acid, etc., for oxidations, and zinc, stannous chloride, iron, etc., for reductions when acids or alkalis are present. Ion discharges, forces that are ultimately obtained from an electrical energy source, such as coal or waterpower, are the only mechanism responsible for producing this phenomenon. The actual material consumption is quite minimal in these processes, even in electrolysis, where the presence of specific chemicals with distinctive oxidizing and reducing capabilities is an essential component of the reaction. Since materials like metallic salts are regenerated by the current after serving their purpose and can be utilized to carry out innumerable reactions, they only need to be present in the electrolyte in extremely minute amounts. Only the issue of energy needs to be taken into account in this instance.

The electrochemical approach enables the reaction to be contained inside a specific area of the chemical system. Only the immediate area around the electrode experiences the reaction; therefore, the reactions of the ions themselves occur on the electrode surface at the time of their discharge, while those of the depolarizers occur in proportion to the quantity coming into contact with the electrode surface, either through diffusion or stirring. The size of the area where the reaction takes place thus relies on the size of the electrode surface; it can be thought of as a very thin layer that is near the electrode. The speed is determined by the concentration of the active molecules. But these are the ions that were just released, either on their own, when they interact with one another, or simultaneously with the depolarizer molecules. Faraday's law states that the concentration of the former is unaffected by the electrical conditions, whereas the concentration of the ions is determined by the strength of the current. The size of the reaction area and the number of reactive ions present inside constitute the electrically feasible reaction conditions, meaning that the concentration of the ions can be controlled entirely electrically and within the ranges. With weak currents and big electrode surfaces, the highest concentrations can be achieved just as well as the largest dilutions with strong currents and small surfaces. These concentrations consequently have a clear impact on the reaction velocity, which is the most crucial component of reaction kinetics.

6.4. ELECTROLYSIS

Electrolysis (Figure 6.1) has one reaction possibility that is always possible. That is the liberation of the ions from their molecular state on the electrode. It is important to distinguish between this reaction and the discharge that came before it. Since the discharged ions which are either atoms or unsaturated groups formed by dissociation cannot exist, it occurs by Faraday's rule, and as a result, they react with unknown velocity. By Faraday's rule, the stable end products are released after they join to form molecules or complexes, with the quantity separated being proportional to the discharge. However, if a depolarizer is present, the discharged ions have a chance to interact with it as opposed to being released. A specific velocity is also used to describe the depolarization reaction. The ratio between an ionic release and a reaction with the depolarizer, however, depends on the velocities. This is where reaction velocities in electrolytic processes come into play. These conditions can only provide information about how electrolysis will proceed quantitatively. The chemical forces of affinity that are unique to the individual elements or compounds and distinctive of the reacting masses influence the course of the reaction.

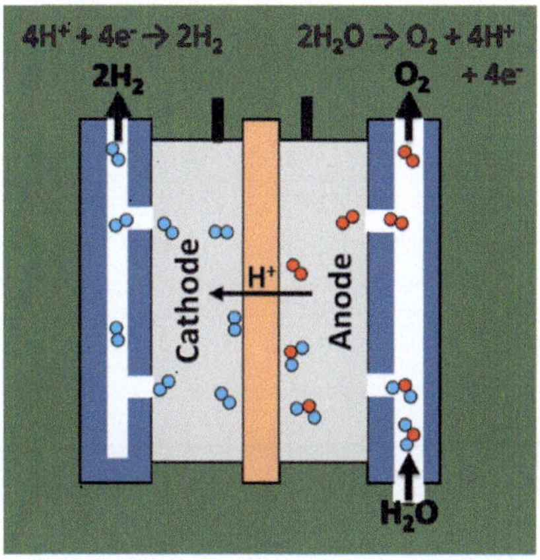

Figure 6.1. Electrolysis.

Source: https://www.energy.gov/sites/default/files/styles/full_article_width/public/2021-07/hfto-electrolyzer-illustration.png?itok=s8WlBKKg, Creative Commons License.

The fact that there are only two possible types of reaction, namely reduction, and oxidation, greatly simplifies the state in the majority of electrolysis of organic molecules. For instance, in the reduction of nitrobenzene, only nitroso benzene, phenylhydroxylamine, and aniline need to be taken into account, and the chemical composition of the reducing ions cannot expand these limits. The efficacy of the reducing agents manifests itself as a quantitative entity that can be repeated because the single reduction phases are quantitatively related to one another, with the succeeding one always being the direct reduction product of the preceding one, and because the attainable phase solely depends on the more or less strong reduction. The individual qualities of the reducing agents are quantitatively comparable to one another. When nitrobenzene is reduced to aniline using copper and sodium hydrate, but only to azobenzene when reduced with zinc and sodium hydrate solution, the specific action of the copper and zinc is demonstrated qualitatively, but there is also a quantitative connection because copper is a stronger reducing agent than zinc, making it a qualitatively equivalent agent. Electrical methods can also produce the products by selecting an appropriate reducing agent. Finding a consistent way to evaluate the effects while simultaneously converting qualitative events into quantitative ones is an important challenge in electrolysis. The current controls the concentration conditions that are electrically possible for the discharged ions, and as a result, it also determines the reaction velocity.

6.5. IONS

The anions move to the anode while the cations move to the cathode when the ions move to the electrodes. This happens as soon as they are close enough to the electrodes for an electrical neutralization to take place. We can assume that this phenomenon occurs on the boundary between the metal and the solution in a way that the ions touch the electrode without actually being on the electrode; the discharge of the ions will take place in a thin layer above the electrode surface. Complex ions produce highly reactive groups that are unsaturated, have free valencies, and are therefore likely to undergo successive reactions (Kehrer, 1993). In elementary ions, this discharge process results in free elementary atoms with high affinity. The behavior of cathodes demonstrates that the electrode cannot experience a discharge. If the ions discharge themselves onto the metal surface, it is quite hard to retrieve. The separation of the electrode will be in two phases: the molecular separation of the discharged ions and the ionic discharge, which transfers ions into an unsaturated state. In general, it happens so quickly that discharge and separation seem to occur at the same time. Following the establishment of an equilibrium between the discharged ions, the atoms or unsaturated groups, and the separation products, the discharge and separation both follow Faraday's law.

Only those cathode processes that involve the reduction of an organic depolarizer are important in organic chemistry. The ions released at the cathode perform this reduction. These ions' chemical composition might vary greatly, and thus, the way the reduction takes place, too. With the assumption that the depolarizer is a nonelectrolyte, hydrogen ions will form in acidic solutions and alkali ions in alkaline solutions. By adding the right substances, any desired type of ions can be activated at the cathode, allowing any metal ions to be released. If the cathode is weak, the metal ions are either generated by the cathode metal itself and pass from this into the electrolyte, or they are added directly to the electrolyte in the form of a metallic salt or hydroxide. If the cathode metal is used and a differentiation is made between reactive and dormant cathodes, the reduction reactions can be achieved easily. Reactive cathodes are those that release minute amounts of ions into the electrolyte either when the current is flowing or not. Naturally, the only reactive cathodes that are of interest here are those that can produce reducing ions. If attackable metal ions are introduced, the metal is not deposited on a non-attackable electrode as long as there are enough depolarizers, and the depolarization velocity is high enough compared to the discharge velocity. Although the cathode metal never changes, reactions occur that can be comparable to metal in the electrolyte. The separation of these metal ions in the electrolyte, and not the hydrogen atoms, is what causes the reducing action on or within the electrode. Thus, we may compare the hydrogen reduction at unreactive electrodes with that of reactive electrodes and the metal ions in the electrolyte at electrodes. We can assume that the hydrogen atoms release themselves onto the surface of the cathode and that the released ions have two options. Either they diminish the depolarizer, or they separate on the cathode. The ratio of division is determined by the reduction velocities of both processes. Practically no hydrogen will be evolved if the reduction occurs much faster than the formation of hydrogen molecules. Thus, when using the current for reduction, the rate of hydrogen production is crucial. It heavily depends on the cathode surface quality and chemical composition, and it probably has something to do with the metal catalytic properties. For the relationships in the electrolyte to be qualitatively the same as when metal ions are injected directly into the electrolyte, ions are released from cathodes. Since the metal used as the cathode hasn't reacted with the electrolyte, it is therefore irrelevant to the effect and can frequently be replaced with a cathode. The metal ions act as hydrogen atoms in these conditions. Depending on their reaction velocities, they discharge in the cathode boundary surface and have an impact on the metallic deposition and depolarizer reduction. As a result, with a high reduction velocity, as long as there is enough depolarizer present, no metal of any kind is deposited on the cathode.

The amount of effort required to change a gram-equivalent of hydrogen from the ionic state to the molecular state at atmospheric pressure in a specific electrolyte is constant and depends only on the initial and final conditions. However, when the same amount of electricity and a gram equivalent of hydrogen are mixed, the electrical work at various electrodes varies, as does the potential for hydrogen evolution with various metals. Since the initial and final states are equal, it follows that any excess electrical work must be offset by a gain in work of the same magnitude. The excess potential is the difference between a hydrogen electrode's potential value in the electrolyte and the discharge potential of hydrogen. A significant amount of heat can be generated by the surplus potential. At metals with excess potential, the reaction rate of molecule formation from the discharged hydrogen ions is slowed down, favoring the reduction of the polarizer over molecular formation. The high potential and excess voltage demonstrate this slowing of hydrogen formation or production. The reduction of materials that are reducible with effort/energy can occur with a high voltage in the discharging space, which is not possible with platinum electrodes but is possible with zinc or mercury cathodes. The concept of excess potential can be used in ion separation at cathodes. Since the possibility of discharged ions being catalytically affected to form stable molecules is always there, the separation potential of every ion likely changes with the nature of the electrode. It has been demonstrated by E. Miller and Cohen that anodic processes also experience excess potential phenomena (Pinaud et al., 2013). Hydrogen does not play a role in metals that provide reducing ions, and each cation is capable of reducing hydrogen. The excess potential of hydrogen cannot be used to explain the distinct reducing properties of copper, zinc, tin, and lead cathodes. The electrolyte plays a role in attack ability; therefore, the rule of excess voltage may apply in one situation but not in another. The excess potential of lead, for instance, can be used to explain the strong reducing action of a lead cathode in sulfuric acid solution, yet the same metal is attackable in alkaline electrolytes and yields reducing ions, whereby the hydrogen action seems excluded. Depending on the cathode material, it appears that the reducing action of compounds differs at cathodes. Some reactions are dependent on the concentration of the reacting substances, which is controlled by the current strength, the chemical resistance of the system, or more specifically, the nature of the medium, and by catalytic reactions that can be attributed to the electrode metal. The cathode potential that was determined for the pure electrolyte is effectively lowered by substances that are difficult to reduce and have a negligible polarizing value. The amount of hydrogen consumed by the depolarizer in the case of such substances, reduced in acid solution at reactive cathodes, is never equal to the amount of hydrogen released by the current. If we assume that the electrode material has a catalytic effect on the velocity of molecular hydrogen formation, the excess potential of the electrode in pure acid without the depolarizer will be significant for the reduction energy, provided that the catalytic property is not altered by the depolarizer.

6.6. CATHODE

Cathode voltage in pure acid affects how quickly caffeine is reduced at lead cathodes. When the excess potential in pure acid is higher, the reduction occurs more readily (Bancroft & George, 1930). Caffeine is metabolized more quickly at mercury electrodes than at lead electrodes, even though both metals have the same cathode potential, or excess potential in pure acid, respectively. The presence of caffeine molecules can also slow down the evolution of hydrogen molecules because caffeine molecules enhance the chemical resistance of this reaction at the electrode surface. In this scenario, both the cathode potential and the excess potential rise. Caffeine's depolarizing effect simultaneously aims to reduce the cathode potential. Although caffeine has been added, a rise in potential is initially noticed if the first reaction with minute amounts of caffeine is high. Lead cathodes experience this phenomenon, however, mercury cathodes do not. Only caffeine's depolarizing effect is visible; this may be because lead and mercury's surfaces have different properties. The amount of hydrogen that is electrolytically evolved at cathodes in a unit of time, or the electrolytic osmotic pressure of the cathode, determines the cathode potential. The amount of hydrogen utilized by the depolarizer does not affect electrolytic osmotic pressure. The concentration of hydrogen ions in the electrolyte determines the potential. It is clear that the modification of the excess potential caused by the depolarizer, the catalytic effect of the metal on the velocity of both the molecular formation of hydrogen and the reaction between hydrogen and the depolarizer, that the cathode potential is conditioned by several reactions that can be independent of one another. As a result, the impact can vary even at electrodes with equal potentials, as is the case, for instance, with caffeine at lead and mercury cathodes.

Compared to cathodic processes, this aspect of the electrolysis of organic molecules is less understood. The activity of oxygen is comparable to that of hydrogen. The excess potential phenomenon also occurs with the oxidation phenomena. As a result, at lead-peroxide anodes, p-nitrotoluene can be converted into p-nitrobenzoic acid, whereas at platinum anodes, only alcohol is produced. The anode can catalytically influence both the oxidation process and the production of molecular oxygen. If the first process is fast while the second is slowed down, we get the excess potential, which causes oxygen evolution to only occur at a high potential. Compared to reductions, electrolytic oxidations are more complicated due to the wide range of ionic activities that could occur. Every ion with the ability to substitute can enter the reactive state at the anode and cause reactions that cannot be counted as actual oxidations. As in the electrolytic synthesis of iodoform from alcohol, substitution and oxidation reactions frequently take place simultaneously. In anodic processes, more distinct types of reactions must be taken into account. Reduction processes are more prevalent than electrical oxidations. The oxidation potential with which oxygen attacks the depolarizer determines the activity of an oxidizer. The benefit of this approach is that it protects oxidizing acids. Thus, this oxidation is both a secondary and chemical process. Nevertheless, the fact that the material is replaced by energy gives it characteristics of an electrochemical reaction. Lead peroxide acts as an oxidant when applied superficially to a lead anode, then it is reduced but always regenerates due to the current.

Hydrogen peroxide is converted to water and oxygen through electrical oxidation at platinum anodes. Electrolytic processes are particularly suited for studying the laws of reaction, especially those of reaction velocities, because the reaction conditions may be easily controlled. According to Faraday's laws, the reacting substances are dependent on the current strength, making it possible to freely alter the overall course of a reaction within large limits. This possibility is less likely to occur in pure chemical processes when changing the conditions of pressure and temperature. There have been attempts to analyze electrical processes from a reaction-kinetic perspective and use them directly to calculate reaction velocities. These processes can be called reactions in heterogeneous systems because the reaction space of electrolytic processes consists of a very thin layer in contact with the electrode. The reaction velocity is influenced by the diffusion velocity or the speed at which the mobile components approach the phase boundary. In diffusion theory (Nernst), the electrical speed of reaction is directly a speed of diffusion; Haber's formula is valid only when the diffusion is minimized (Weingärtner, 2008). The experiment's conditions will determine how the reaction speeds are measured. If the reaction between the two components occurs instantly, we can completely alter the reaction's time at will by varying the amount of time we add one of the components. The measured reaction velocity must, of course, always remain proportional to the additional quantity of the reaction components if the latter is consumed. According to the Nernst theory, the reactive agent supplied by the current is maintained at the electrode by the potential relations in an infinitesimal concentration as compared to the external concentration.

The first inference made is that transference and diffusion determine the current strength because the reaction can only advance if new quantities of the agent reach the electrode. Only through ions can the current get to the electrode. Since the current strength controls the quantifiable velocity of the reaction, it follows that the effects of diffusion and transference also have an impact on this velocity of the reaction. According to Haber's theory, the ions and depolarizers are present at the electrode in higher concentrations than the ions produced by the current strength. Nernst's theories only hold true if the depolarizer creates no concentration along the line of reaction. Because in this scenario, the reaction can only advance in proportion to the depolarizer diffusional entry into the reaction layer. Because the principles governing the reaction can practically be used in heterogeneous systems, even if the entire reaction occurs in the fluid phase, or in a homogeneous system. The layer in which the discharged ions are present may be thought of as a thin film that behaves as a heterogeneous formation toward the electrolyte in the localization of the discharging space in the immediate proximity of the electrode. Given the high rate of reaction or separation, the concentration of the released ions is very low. Therefore, the rate at which additional ions are transferred to this layer by diffusion and transference determines how quickly the reaction will proceed.

6.7. CONCENTRATION

The concentration of the discharged ions can be governed by the laws of gases. According to Haber, the discharged ions must have an osmotic pressure proportional to the concentration in the discharging space, or the current strength, as they are not in or on the electrode but rather in the solution (Yang Zhao, Song, & Duan, 2022). When ions are assumed to be dissolved in the electrode, it is challenging to understand how the laws of gases can still apply, especially when metal ions enter the cathode and cause reduction effects. The formal validity of the Haber assumption is intact, but its scope of application seems to have increased because cathode behavior is now theoretically proved under the constrained conditions. In addition to encouraging more research, the multiple challenges and several points support the idea that organic electrolysis is ideally suited to transferring physical chemistry principles into the field of organic chemistry. It is crucial to have a setup that enables the observation of the critical potentials as well as their control. Important elements include specific electrolyzing equipment fit for a specific purpose, which frequently has a significant impact on an experiment. The conductivity of the electrolyte, the necessity for gas collection, the separation of the cathode and anode chambers, temperature control, and variation in electrode size call for specific setups and requirements.

Immersing the two electrodes in a glass vessel while maintaining their alignment, typically parallel to one another is sufficient. When collecting gases, the vessel is sealed with a stopper. A glass tube must fit through one of the stopper perforations, and the electrodes must fit. Mercury helps to complete the circuit. By heating or cooling the vessel from the outside, temperature changes can be achieved. The conductivity can be changed by choosing different concentrations, as well as by changing the voltage. The cell electromotive force defines the current strength; its relationship to the electrode surfaces determines the current density; and its relationship to the electrolyte volume determines the current concentrations. The cathode and anode spaces must typically be divided for this design to work. The easiest way to achieve this is by using a diaphragm, but it can also be done by connecting two distinct vessels, one containing the anode and the other the cathode fluid. Typically, porous cylinders or plates serve as diaphragms. Sometimes gypsum, crushed asbestos, etc., are used to make diaphragms. Also employed are so-called "acid-proof" diaphragms. Plates are simply fitted securely or glued into cylindrical vessels after they have been placed in the outer vessel. Making the cell of two distinct sections fit upon one another is a simpler technique. The diaphragm plate is firmly wedged between these using screws. In electrolytic investigations, cooling and stirring are very crucial. A constant temperature of the electrolyte can be achieved in addition to external cooling by inserting glass or porcelain through which a constant water current is carried. Only if they are to function as electrodes at the same time must metallic sheets/rods be employed. If not, they behave rarely as intermediate conductors during the electrolysis. Since their surfaces restrict the reaction area, it is frequently important to keep the electrodes cool.

The cathode or anode potential of the electrode at which a specific reaction occurs has an impact on how electrolysis will proceed. On the other hand, the potential difference between the electrodes, which is affected by a variety of factors, including the resistance of the diaphragm, etc., is typically of no significance for the reaction. A wire connects a secondary or standard electrode to the liquid surrounding the experimental electrode, whose potential is either arbitrarily assumed to be zero or has a specific absolute value. The electromotive force is then assessed using a galvanometer or capillary electrometer, one of the widely used techniques. The difference in potential of the reaction electrode, based on the established zero value of potential, is derived by accurately subtracting the potential difference of the standard electrode from the result. The calomel electrode and the hydrogen electrode of Nernst are the two secondary or standard electrodes in use. It consists of a combination of mercury covered with mercurous chloride as a depolarizer and submerged in a solution of potassium chloride solution. The hydrogen electrode is made of a platinum sheet that has been charged with hydrogen and is submerged in an electrolyte. To create the hydrogen electrode, sheet platinum (or palladium) is positioned so that it is partially submerged in the electrolyte and partially submerged in hydrogen gas. A steady current of hydrogen is then sent through the electrolyte to maintain the saturated state. Thus, a hydrogen environment must be maintained around the portion of the electrode that is not submerged in the electrolyte throughout the duration of the experiment.

6.8. ELECTROLYTE

The electrolyte of the normal electrode must not react with that of the experimental cell, and in the majority of cases, it will be advantageous to separate the two using a suitable diaphragm (Brattain & Garrett, 1955). The measuring of the electrode potential gives rise to a variety of issues. Finding the potential at which a reaction starts, or the discharge potential that the reacting or separated ion possesses in the presence of the depolarizer, is a common task. The decomposition potential can be measured to determine this value most easily. This technique is based on the idea that an electrolyte can only permanently decompose by using an electromotive force that just manages to outweigh the polarization. When we polarize a cell with a weak electromotive force, the current initially cannot leave the cell permanently. The sudden deflection of a galvanometer enclosed in the circuit only occurs when the electromotive force exceeds the value of the polarization and reveals the electrolyte's decomposition value. When the electromotive forces, current intensities, or deflection factors of the galvanometer are viewed as factors in a coordinate system, curves are produced that exhibit distinctive peaks at the decomposition values. If the electrode where the expected reaction will take place is combined with a regular electrode, and the difference in potential at the work-electrode is observed for increasing current strengths, it will be revealed that at a particular value of the latter, a sudden passage of the current, which manifests as a break in the curve, occurs. Any form of reaction, whether it is the separation of ions or their reaction with the depolarizer, will typically start with this split. These breaks in the curve can occur repeatedly when different types of ions are split apart or react with various electromotive forces. Measuring the discharge potential of the cations or anions before and after the addition of the depolarizer that is to be acted upon is the simplest way to determine whether a reaction will occur or not. The commencement of the reaction is indicated by a change in potential.

Knowing the probable window of time within which one or more specific reactions can occur is especially crucial. The answer to this question is determined by the difference in potential that an electrolyte with a depolarizer creates compared to one without one. The chlorine ion decomposition in combination with the hydrogen electrode is found at 1.31 volts in a n-hydrochloric acid solution. If phenol is added to the solution, the curve breaks at a voltage of 0.9 volts. In light of this, the range of potential within which the reaction for the synthesis of phenol-chlorine derivatives must occur is 0.9 to 1.3 volts. This method was used to assess the OH ion's ability to decompose when combined with a hydrogen electrode in a diluted sulfuric acid solution, both with and without the addition of ethyl alcohol. The reaction between hydroxyl ions and alcohol might result in the formation of either acetaldehyde or acetic acid. The hydroxyl ion decomposition potential was measured in diluted sulfuric acid and in the presence of acetaldehyde, but there was no decrease in the potential. Given the current conditions, the acetaldehyde does not serve as a depolarizer, therefore if the electrolysis potential is maintained between 1.3 and 1.6 volts, quantitative yield of acetaldehyde must be obtained. These theoretical inferences were fully supported by the experiments. The main issue is to maintain this potential at a specific level or within specific bounds. This is done by choosing appropriate polarizing current strength and setting the cell, which consists of the work-electrode and the normal electrodes. It is also done by making sure that the tension between the work electrode and the standard electrode maintains the value of the compensating potential by varying the current strength as may be necessary throughout the experiment. The measurement of single electrode potentials, or the potential differences of an electrolyte in connection with a specific electrode with or without a depolarizer, is crucial in and of itself for obtaining the depolarizing values, as we've discussed before. It is clear that these depolarizer values are defining characteristics of the depolarizer's chemical nature and are intimately connected to the structure and arrangement of the molecule.

For electrolytic processes, the type of the electrode is crucial. The nature of the surface and the previous treatment of the electrodes can have a significant impact on how the electrolysis proceeds. It is obvious that the size of the surface soaked by the electrolyte determines the potential and current density, and even on this basis, its smoothness or roughness form decisive factors. However, it is also important to take into account the form of the surface as well as the mutual position of the two electrodes because these two factors determine how the lines of force are distributed on the surface. Since there are an uneven number of discharging current lines, data on current densities and potentials often relate to mean values even though both are frequently variable at different locations on the surface.

Therefore, smoothing the surface is frequently advised, especially when taking precise potential measurements. The relative size and placement of the two electrodes must be taken into account if the purpose is to get similar current densities.

Concentric cylinders make up these electrodes, between which electrolysis occurs. By initially coating the electrode anodically with a thin layer of lead peroxide and then decreasing this cathodically, such a smooth surface can be easily created (Zhou, Dai, Lei, Ma, & Wang, 2005). The anodic process results in a solution of the foreign metals in the surface coat, and reduction yields a pure lead surface. It is crucial to have a pure cathode using various techniques. Impurities of any size can significantly change the product. The traces of contaminants catalytically influence the ions velocity of separation. In fact, since the impurities lower the reduction energy, we must infer that the hydrogen production process is catalyzed more quickly. There will be a disturbance in reduction if there is an increase in current, or, more precisely, when there is a decrease in the potential difference at the cathode, with a constant-current source and external resistance. The metal is electrolytically coated onto electrodes under the right conditions by electrolyzing a pure salt solution. Thus, we can obtain cathode potentials over an extended period of time even with reactive cathodes. The cathode potential quickly decreases, even if the current is constant, and the evolution of hydrogen stops if powerful currents are continuously passed through the cell for prolonged periods when a depolarizer is present. When the current is restarted after some interval, the initial potential and hydrogen evolution returns. As a result, the electrodes depolarize more effectively with hydrogen than without it. With different metals, this influence can be extended to varying degrees. One can distinguish between organic chemicals that behave as electrolytes and those that just serve as depolarizers when they are degraded in solution by a direct current. Even if anodic and cathodic effects are occasionally seen side by side or successively in electrolysis, the classification into oxidation and reduction processes would also be useful in the presentation of the experimental data.

It has been noticed that the solution forms when carbon is used as the anode in an acid electrolyte, and it has also been noted that carbon frequently appears in the cathode precipitate during galvanic metal deposition. It was discovered that the anode carbon is attacked during the electrolysis of dilute sulfuric acid employing carbon electrodes. This was demonstrated by the appearance of carbon monoxide and dioxide. Under the right conditions, carbon dissolves in solution, coloring the sulfuric acid. As a component of the cation, carbon travels to the cathode and deposits itself as a conductive coating on the platinum cathode, acting like a metal. They appear to be an undesirable material for electrolytic studies due to the high chemical resistance of aliphatic hydrocarbons and the aggregation condition of their carbon-poor constituents. These reactions are cathodic when hydrogen is added, anodic when halogens are added, etc.

CHAPTER

7

Intermolecular Interactions

Contents

7.1. INTRODUCTION

The thermodynamic parameters of the vaporization process, the enthalpy of formation, and their dissociation energies were established by extensive investigation of organometallic and organoelemental compounds in various states. The physicochemical properties of molecules providing information on the features of intramolecular and intermolecular interactions are the thermochemical properties and the dissociation energies of organic complexes. There are numerous discrepancies between the existing theories and the observations from the study of inorganic and organic substances. One of the unique characteristics of these compounds is the availability of the spatial structure in the liquid state with a three-dimensional network structure incorporating chain and ring fragments and the donor-acceptor nature of the intermolecular hydrogen bonding (Figure 7.1).

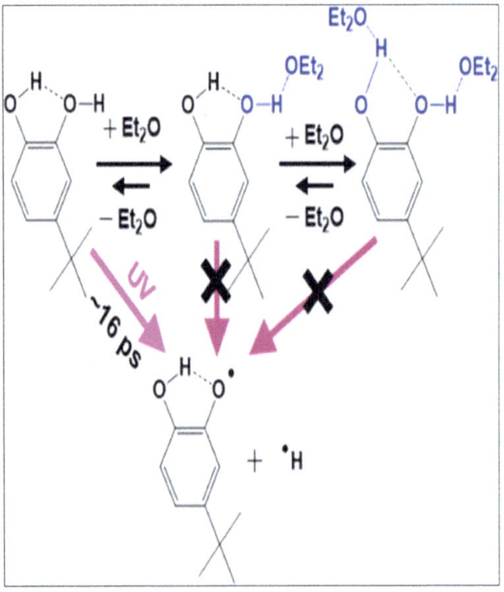

Figure 7.1. Intermolecular hydrogen bonding.

Source: https://pubs.acs.org/cms/10.1021/acs.jpca.9b04573/asset/images/acs.jpca.9b04573.social.jpeg_v03, Creative Commons License.

Different homological series compounds containing the hydroxyl group are distinguished from one another by long intermolecular hydrogen bonds. Understanding the composition of the solvents and how they interact with the complex molecules is becoming more difficult. The thermodynamic parameters of alkyl compounds that have been confirmed through experiments do not match the values predicted by the sp_3-hybridization model of a carbon electron configuration (Zhao et al., 2020). This model is unable to account for the causes of the rise in enthalpy in homological series, such as gallium alkyl compounds.

7.2. VAPORIZATION

The homologous series of saturated and unsaturated hydrocarbons exhibit symmetries. The data on the vaporization enthalpy of different organic compounds show that there is no association between the molecular mass and the enthalpy, energies of the donor-acceptor bonds, etc. The thermodynamic characteristics of the vaporization and sublimation processes are used to estimate the energy of hydrogen bonds and the specific interactions of organic compounds. Organoelemental substances serve as a connecting link between inorganic and organic chemistry. The idea put forth by Sidgewick that compounds with the 18-electron configuration of metal atoms were unable to bond was disproved by the discovery of strong intermolecular interactions in the solid and liquid state as well as by the demonstration of the existence of dimeric molecules in the vapor of nickel, iron, molybdenum, and tungsten carbonyls (Chattaraj & Maiti, 2001). The nature of the bonds in different aggregation states, of specific and nonspecific interactions in water and nonaqueous solutions, is known for the number of inorganic compounds.

The models of sp, sp_2, and sp_3 hybridization revealed how molecules are bonded. However, the sp_3 hybridization model for the carbon atom in alkyl compounds doesn't explain methane volatility. The electron density shifts along the alkyl chain in the alkyl compounds of the elements from the nitrogen and oxygen subgroups about their electronegativity. This can be from the 2s-orbitals of the carbon atoms of the alkyl group. The shifting of the electron density away from the carbon atom of the terminal methyl group is hindered by the interaction of the carbon atoms in the ethyl, propyl, and other alkyl groups. As a result, the adjacent carbon atom experiences the reverse electron transmission, while the terminal carbon of the methyl group in the ethyl and propyl substituent experiences a drop in negative charge and consequent increase in affinity. Three different types of bonds are thought to be present in the formation of the ferrocene compounds with equal Fe-C interatomic distances under the conditions that the p-shell restricts the entry of the ring C_5H_5 into the Fe-C_5H_5 bond. These bonds are covalent and this results from the presence of two unshared pairs of p-electrons, two unoccupied p-orbitals, and one unpaired p-electron in the C_5H_5. As a result, the bond in the alkyls of the main group elements differs significantly from the existing bond of a similar character in the compounds of acetylene and ethylene with transition elements. The number of C atoms in the alkyl group increases. This occurs until there is no impact on the distribution of electrons between the central atom and the carbon atom bonding it. The increase in the number of C atoms in the alkyl group increases and has no impact on the change in the dissociation energy. It must be observed that the presence of an even and an odd carbon atom in the center of an alkyl chain affects the displacement of the electron density in the chain.

Propyl, butyl, and other dissociation energies are not significantly different from one another, making it impossible for us to assume that each bond contributes equally to the dissociation energy. It follows that the dissociation energy is unaffected by the length of an alkyl chain with more than three carbons. As a result, neither the steric effect nor the stability of the compound is impacted by the lengthening of the alkyl carbon chain. The reverse ionic bond weakens as the dissociation energy decreases from the methyl to the ethyl compound, hence the steric effect may only exist between carbon atoms that are in direct contact with the central atom. In other words, it is incorrect to assume that the increasing number of carbon atoms in an alkyl chain causes the dissociation energy to decrease in the series $Zn(CH_3)$-$Zn(n\text{-}C_4H_9)$, $Sn(CH_3)_4$-$S(n\text{-}C_4H_9)$, and $Te(CH_3)_2$-$Te(n\text{-}C_4H_9)$. The important finding of the thermodynamic analysis is the insensitivity of the branched alkyl compounds to the steric effect. Instead, they offer more reliable complexes. This discovery raises doubts about correlation analysis in the chemistry of organoelemental and organic molecules when taking steric effects into account. A thorough examination of the dissociation energy, thermodynamic characteristics, and information on the distribution of electron densities may serve as the foundation to understand the steric effect.

The reverse ionic bond is a vital characteristic of organometallic, organoelemental, and organic compounds (Denmark & Fu, 2003). The reverse ionic bond in alkyls stabilizes the molecules of alkyl compounds and has an impact on all physicochemical characteristics, including their ability to dissolve in solutions, their dissociation energies, and the energies of interactions in a condensed state, nonelectrolyte solutions, and solvation processes. The methyl group of the CH_3OH molecule has valuable properties that are required for the understanding of the nature of interactions in alcohols, functional solvents, ketones, ethers, and carboxylic acids. It possesses an almost entirely unshared $2S_2$ electron pair as well as donor or acceptor properties depending on the interacting functional group in the bonding molecule. The nature of the specific interactions, which are dependent on the bond vacancy of the molecule, functional group, and the CH_3 group possessing the mode of the specific interaction inherent to it, are directly governed by the differences in the structures of the organic compounds and their solutions. The latter comes from the essentially unshared $2S_2$ electron pair in a pseudoatomic system. As a result, several specialized interactions emerge in the liquid state for the functional solvents, particularly N-methyl formamide and this molecule contains different functional groups of varied character. To investigate liquid structures and analyze their thermodynamic and other physicochemical aspects, new approaches must be developed. None of the study techniques now in use offer clear details about the solvent and solution structures. For instance, it is hardly possible to establish the structure of two distinct interactions with close energy values in a solution of a dissimilar character based solely on spectral data or quantum computations.

7.3. STRUCTURE

The examination of the structure encounters challenging issues due to the method's ambiguity. The spatial distribution of atoms and molecules is transformed into a homogeneous function via the radial distribution function (RDF) (Levine, Stone, & Kohlmeyer, 2011). This method limits the use of the concept of a molecule to comprehend the structural arrangement in a liquid, such as when a hydrogen bond forms in alcohol. $H–O–CH_3$, is formed by the unshared $2S_2$ electron pair of the carbon atom from the methyl group in the case of an energy difference. Two distinct configurations, in particular, with a bigger energy gap between the hydrogen bond and the special interaction $H–O–CH_3$, may be characterized by identical RDFs at the same time. This feature limits the application of X-ray diffraction. The understanding of the structure of liquid dimethylformamide, for instance, was expanded by X-ray diffraction research. Therefore, the SP_3 hybridization model of the carbon atom should be omitted in the probabilistic structure of liquid and the existence of the unshared $2S_2$ electron pair forming the specific interaction and providing the carbon atom with the pentacoordinate state should be taken into consideration.

Quantum calculations support the idea that the methyl group provides the characteristic of a functional group with a free bond vacancy as a result of the carbon atom configuration. An example of a dimeric trimethylaluminum molecule shows how this functional group engages in chemical coupling between a different aluminum atom and the carbon atom of the methyl group, and how this bond is localized on the orbitals of the interacting species. The P_z orbital of the aluminum atom is where the pseudoatomic, unshared $2S_2$ electron pair of the carbon of the $Al(CH_3)_3$ involved in the interaction with the aluminum atom of a molecule in the vicinity is located. The interaction is of a donor-acceptor nature and falls under the new category of $AlCH_3$–Al interactions. The concept of intermolecular contacts/bonds is expanded by these kinds of donor-acceptor interactions. The carbon atoms of the methyl groups with the metal in the metal alkyls are formed by the same donor-acceptor association. In liquid alkyl metals, the interactions develop between the molecules of the nearest area or space. The presence of the unshared $2S_2$ electron pair in the alkyl derivatives opens up a wide range of possibilities for the emergence of specific interactions of the same kind with energies and charges differing on the atoms of elements and carbon.

7.4. THERMODYNAMICS

Trimethyl boron B-CH_3 has a negative charge on the carbon methyl group that is 0.234 times more than the charge on a similar group in trimethylamine N-CH_3, which leads to high energy bonds. Thus, the methyl group of trimethyl boron has donor properties due to the pseudoatomic $2S_2$ electron pair. In the binary systems of metal alkyls, these interactions with low stability may occur. The increasing number of functional groups in a molecule, whose bond vacancies create distinct interactions, complicates the structures of liquid organic molecules. The Krestov concept is acceptable for the wide variety of nonelectrolyte solutions. This concept is based on the solution's dominant structural characteristics. Certain compositions are found in the solutions where the solvent's structure is determined by the structure of the pure components or the structure of their mixtures. The precise interactions that are generated in the solvate and the specific interactions that are changing in energy in the liquid component are what provide these solutions their unique features. The interpretation of the observed solution properties as a function of its composition is constrained by this defining feature of nonelectrolyte solutions.

The establishment of the thermodynamic properties of the process allows us to assess the energies of the specific interactions present in the solvate using knowledge of the vapor composition because the solvation is the sum of energy and structural changes in the process of the solvent molecule transition into the liquid phase of the solution. The nonspecific bonds are not taken into account when calculating the energy of specific interactions. Entropy as a structural characteristic refers to the energy difference in the intermolecular interactions originating in the structures of the liquid compounds when the structure of the liquid component and the solvate are compared. While understanding the structure of organic liquid and solid compounds helps us to understand their thermodynamic properties, understanding the thermodynamic features of vaporization enables us to assess the energies of intermolecular interactions. The critical aspect is the ability to assess the energy of the individual bonds using the enthalpy properties of vaporization by accurately estimating the bond vacancies of the molecule. In the case of liquid water with a tetrahedral structure in the crystal and in the liquid state at a temperature near the melting point, the molecule has four free bond vacancies forming four hydrogen bonds. This cleavage at the vaporization allows us to evaluate the energy of the hydrogen bond at 10.99 kJ/mol from the enthalpy characteristic of vaporization. Tetrameric and other water molecules with fewer rings form in the liquid state of water at higher temperatures, hence the hydrogen bond energy has an average value. Given that monomer molecules are dispersing into the vapor, the decrease in the number of molecules causes a minor stabilization of the hydrogen bond, which is accompanied by a rise in temperature. Even so, the differences in its energy, as measured at the non-standard temperature, result from variations in thermal motion. The nonspecific interactions are not taken into account when calculating the energies of the specific interactions using thermodynamics, and the hydrogen bonds used to calculate the energies of nonspecific interactions have limits that are as low as those used in experimental measurements of the enthalpy characteristics.

Given that the elementary volume of a liquid is useful when it contains the lowest number of species, it is possible to estimate the energy of a given interaction by knowing the quantity and type of bond vacancies among interacting species. The calculation of the energy of the hydrogen bond in liquid water needed data on the number of bond vacancies, the enthalpy of vaporization, and the composition of the vapor. A system of self-consistent energies of the particular interactions in molecular liquids is produced as a result of this method. Here, taking into account the molecule structure and any bond vacancies is crucial. Fundamentally, this method is already used with functional solvents, alcohols, and carboxylic acids. However, the sole factor considered for each of these compounds was the quantity of hydroxyl or carboxyl groups. Only the OH groups were taken into account in the quantum-chemical calculations for the evaluation of the energy of the hydrogen bonds. Therefore, we can use some methods to estimate the energy of undiscovered types of specific interactions in the molecular liquids of organoelemental compounds. The zinc alkyl metals have a linear shape. The planar triangular structure of the boron alkyl metals is present. Investigations using X-ray, electron, and IR spectroscopy as well as the diffraction of light revealed the existence of monomer molecules known as gallium and indium alkyls in vapor. Trimethylaluminum is present in vapor in the form of dimers, triethyl aluminum is less dimerized in vapor, and $Al_2(n–C_3H_7)_6$ has lower partial pressure in vapor. Trimethyl and triethyl aluminum was studied in vapor using X-ray, electron, and IR spectroscopy (Kim, Ho, Suh, Prakash, & Bunshah, 1991). Both in the liquid and solid states, alkyl aluminum can be found as dimers. According to electron diffraction experiments, trimethylindium is a monomer in vapor and a tetramer in liquid when it is in a crystal. It was determined through X-ray diffraction analysis of the crystalline trimethylindium that a group of four indium atoms and trigonal units, designated $In(CH_3)_3$, existed and that they were almost exactly situated in a plane parallel to the plane of In_4. The methyl groups sandwiched between the indium atoms hold the tetrameric unit together. The carbon atom is consequently in the pentacoordinate state.

The study of the structure of the trimethylindium crystal provided the structural parameters and interatomic distances. Consequently, five pentacoordinate carbon atoms are present in the trigonal bipyramidal structure of the $In(CH_3)_3$ crystal. Trimethylgallium can also form crystals with a similar structure. Trimethylgallium spectrum research also revealed the presence of dimeric molecules similar to those of $Al_2(CH_3)_6$ in the liquid state. However, the liquid trimethylgallium was not discovered to contain dimers after this chemical was examined using X-ray diffraction and IR spectroscopy. Trimethylaluminum exists in the dimeric state, which leads to the conclusion that this molecule contains four bond vacancies, and the aluminum atom coordination number is also four. The number of coordinated molecules increases in the series of alkyl compounds of the aluminum subgroup elements, and in the liquid and crystal states of the series gallium-thallium alkyl compounds, five molecules are coordinated around the central bipyramidal structure. When in contact with the methyl groups of other molecules, the metal atom in alkyls of the zinc subgroup MR_2 of the molecules can pick up an additional two electron pairs. The metal atoms (Zn, Cd, and Hg) and the molecules of these alkyl derivatives interact specifically through two M-CH_3 interactions and two CH_3 groups. The methyl groups create alkyl derivatives of elements in the fourth through sixth groups of the periodic table that correspond to the subgroups of carbon, nitrogen, and sulfur. The alkyls of the sulfur subgroup elements with two unshared electron pairs on the atom of the core element create the structure of the liquid state resembling that of water by exploiting four bond vacancies.

7.5. NITROGEN

The nitrogen alkyls have a similar structure. All of the hydrogen atoms in ammonia (NH_3) and the carbon atoms in the CH_3 groups of the alkyl derivative participate in the formation of hydrogen bonds. There are six coordinated hydrogen atoms or methyl groups located around the central atom. Since all bond vacancies should be filled, the nitrogen atoms typically form pentacoordinate bonds, and these elements are known to form complexes with octahedral structures. The central atoms of the alkyls of the second to fourth main group elements have positive charges and exhibit acceptor characteristics in the development of the specific bonds. The atoms of the fifth to sixth main group elements have negative charges and display donor qualities in the development of the specific interactions. The pentacoordinate carbon atom participates in interactions that are common to all alkyl compounds of the major group elements. The calculated energy of the specific interactions was based on the known bond vacancies of the major group alkyl compounds. According to calculations made using the vaporization enthalpy, the energy of specific intermolecular interactions can be higher than the energy of hydrogen bonds in liquid water (10.99 kJ/mol) and are essentially higher than the energy of the same type of interaction in liquid ammonia (3.89 kJ/mol). Low stability nonspecific interactions were ignored in the computations because their values did not exceed the limit in the experimentally measured thermodynamic properties. The methyl and methyl-ethyl derivatives maintain the regularity of energy fluctuation of the specific interactions in the series of the main group elements. The low errors in the experimentally measured vaporization enthalpies of the second, third, and fifth group element alkyls indicate that the energy variation of the specific interactions in alkyl cadmium, alkyl gallium, and alkyl phosphorus deviates significantly.

The donor characteristics of cadmium and gallium atoms differ as a result of the positioning of differing quantities of electrons on various energy levels (Gorczyca, Svane, & Christensen, 1997). The phosphorus atom has acceptor characteristics in its interaction with the carbon atom due to the presence of five electrons on its third energy level. As a result, the carbon atom exhibits acceptor properties with the zinc and gallium atoms and donor properties with the phosphorus atom. This is because the carbon atom has an essentially unshared $2S_2$ electron pair. The fact that the X-ray data-based comparative analysis of the electronic structures of the molecules of $B(CH_3)_3$ and $Al(CH_3)_3$ reveals an increased positive charge on the aluminum atom and a reduced contribution of the Al_3 P_z-orbital to the P-molecular orbital (MO). A distinct sequence of the 3D levels is seen in the zinc dimethyl mercury. Because Hg-CH_3-Hg has a low value (8.32 kJ/mol), dimethyl mercury differs from the standard series in the energy of certain interactions. All of the main group alkyl compounds adhere to the general rule. The energy of these interactions increases as the alkyl chain carbon atom number increases. This growth, however, cannot continue indefinitely, and as we add more atoms to the chain, the energy difference of the intermolecular interactions decreases in each successive step. On the zinc and gallium alkyls, it has been demonstrated. However, the difference between compounds with even and odd numbers of carbon atoms in the alkyl chain and between compounds with odd and even numbers of carbon atoms in the alkyl chain is always smaller. The M-C bond dissociation energy follows a similar pattern in the series of alkyl derivatives. This effect is caused by differences in the reverse bond that affects the energy of specific interactions. This means that the even or odd number of carbon atoms in the chain largely determines the shift of the electron density along the chain from the central atom to the terminal methyl group in compounds of the elements from the subgroups of zinc and boron and along the carbon-carbon chain to the central atom of the fifth and sixth groups in the complex. Therefore, the number of atoms in the chain determines how the reverse bond between the central atom of the alkyl compound and the adjacent carbon atom behaves.

7.6. BOND

The influence of the reverse dative bond can be obvious by the steep decline of difference in the vaporization enthalpies at three carbon atoms in the chain, and on the other hand, by increasing energy of the specific interactions. It is indicative that the charge on the mercury atom in the dimethylmercury molecule (+0.023) changes to negative in $Hg(C_4H_9)_2$. It should be noted that the phosphorus atomic charge changed from PH_3 (0.04) to $P(CH_3)_3$ (+0.070), reflecting the effect of the reverse bond when the hydrogen in these compounds was replaced by the methyl group. These results demonstrate that the reverse bond in the alkyls of the second to fourth-group elements is opposite from the fifth and sixth-group elements. The significance of the reverse dative bond diminishes when the methyl group is replaced by the ethyl and then the propyl groups. All alkyl compounds share the pentacoordinate carbon atom in these interactions. Understanding the structure of a molecule and the number of free bond vacancies is necessary to know the structure of the liquid state of an organic compound and the types of specific interactions. The enthalpy (entropy) of distinct phase transitions is related to the energy and the number of specific interactions, which in turn makes it possible to determine the energy of the latter with knowledge of the thermodynamics of vaporization. Therefore, the molecular mass cannot determine the thermodynamic properties of phase transitions, and even in the homologous series, it is important to take into account the unique characteristics of intermolecular interactions. However, there are problems observed even with alkanes.

Triclinic is the typical crystal structure for hydrocarbons with an even number of atoms, such as n-C_6-n-C_{20}. Polymorphous transitions occur in compounds with a lot of carbon atoms in the chain. Triclinic crystals, as well as the monoclinic modification for compounds n-C_{26}-n-C_{36}, are characteristics of their low-temperature phase in alkanes of the sequence n-C_{20}-n-C_{26}. The high-temperature phase has a hexagonal structure and is stable up to the melting point. From n-C_9 through n-C_{35}, alkanes having an odd number of carbon atoms go through polymorphic transformations. At low temperatures, they have a rhombic structure, while at high temperatures they are hexagonal. Two convergent curves that correspond to compounds with even and odd numbers of C atoms in the chain should also be used to explain the vaporization enthalpies of the saturated alcohols of the usual structure. The essentially unshared $2s_2$-electron pair of the carbon atom involved in the distribution of the electron density controls this aspect of the characteristics of alkanes. The properties of the proton in the hydroxyl group trans-position in relation to the hydrogen are distinct from those of the other two protons in the methyl group. A rough estimate of the energy barrier preventing the rotation of the CH_3-group in the gas phase is 4 kJ/mol. The concept that methanol has a high barrier to rotation in its condensed phase is incorrect, though, as the energy of each connection in liquid and solid phases reduces due to the coordination effect and a greater volume of interactions with molecules in the immediate vicinity. Therefore, the energy of the bonds of low stability under consideration should be lower than 4 kJ/mol for liquid and solid methanol with its characteristic hydrogen bonds and ability to dimerize in vapor. This sort of interaction is 10 times stronger than the nonspecific interactions, according to a thermodynamic analysis of the hydrogen bonds in organic compounds in the liquid state. Understanding their coordination ability, which gives information on the number of bond vacancies, types of hydrogen bonds, and specific interactions, is a key issue for the thermodynamic investigation of formamide, N-methyl formamide, and N, N-dimethylformamide. The capacity of these chemicals to form complexes with hydrogen fluoride and hydrogen chloride is particularly interesting in this context.

7.7. COORDINATION MOLECULES

The reciprocal positions of the molecules determine the large range of the energy of H-bonds in complexes $(HF)_m$ DMF at the mixing enthalpy of initial components, which is between 20.9 and 188.4 kJ/mol. The complexes with a single molecule of dimethylformamide are the most stable among those with an equal number of molecules. The largest energy gain occurs when more HF molecules are added to it. Studies combining IR spectroscopy, quantum-chemical computations, and densitometry have produced results that are very consistent. The results obtained are comparable when these investigations are compared to those carried out using spectral methods and calculations for the system HCl-DMF. Conclusions can be drawn from these data that DMF only forms stable heterocomplexes with hydrogen fluoride and chloride in the concentration ranges where its coordination numbers are 4, 5, and 6. As a result, the two oxygen unshared electron pairs have two bond vacancies and interact with molecules of hydrogen fluoride or chloride to produce the stable hydrogen bonds O-H-F. This demonstrates the improved stability of the specific interaction formed by the essentially unshared $2S_2$-electron pair of the carbon atom of the methyl group and the unshared electron pair of the oxygen atom $N\text{-}CH_3\ O_{14}C$. The methyl group and the unshared electron pair of the nitrogen atom produce the second type of specific interaction, which is less stable. There are three different types of six hydrogen bonds that are connected to a network by interactions with varying degrees of stability.

Since the number of coordinated molecules in the dimer is reduced from six to two and the number of intermolecular bonds is reduced from six to two, the energy of the hydrogen bonds in the liquid formamide does not fundamentally conflict with the presence of dimers in its vapor. According to the comparable relationship between the vaporization enthalpy and dissociation energy of the triethyl aluminum dimer, the fulfillment of these criteria is required for the dimerization to occur in the vapor. Acetonitrile does not form hydrogen bonds and is located in a liquid environment with two molecules facing in opposite directions. This pattern suggests the development of poorly bonded dimers with minimal stability and a vaporization enthalpy of 32.7 kJ/mol. The specificity of its intermolecular interactions in the liquid state is not understood yet. Applying this theory, we can propose that the acetonitrile molecule contains two bond vacancies coming from the N unshared electron pair and the CH_3 atom's unshared $2S_2$-electron pair, resulting in chain and ring configurations in the liquid acetonitrile. These particular bonds have an energy that is four times greater than the interaction DN-CH_3 N (4.20 kJ/mol) present in liquid trimethylamine and are 1.5 times more stable than hydrogen bonds in water. It is more reasonable to think that the nitrogen atom's unshared electron pair and the carbon atom's triple-bonded $2S_2$-electrons combine to generate more stable specific interactions. Similar to methanol, the residual bond vacancy of the carbon atom from the methyl group interacts with the hydrogen of the methyl group in the next molecule to generate H_3CH-CH_2. The schematic representation of the acetonitrile's liquid state is well supported by the RDF of the substance. Thus, the four-bond vacancy acetonitrile molecule generates two bonds, –CN → CN of high stability, and two particular interactions, $DH_3C \rightarrow$ H–CH_2 of low stability, both of which have an estimated energy of 0.60 kJ/mol. The difference between the vaporization enthalpy and the overall energy contribution of the interactions is used to determine the energy of the interactions D-CN→CN on the complex formation with different solvents, the conclusions on the six-coordination state of N, N-dimethylformamide, N-methylformamide, and formamide can be strengthened (Simm, Türtscher, & Reiher, 2020). It should be noted that no precise equation can accurately describe how the qualities of the different components lead to the ideal solution viscosity. The viscosity isotherms of some systems are described by semi empirical equations that are not well suited to describe the viscosity isotherms of other systems.

The experimental results on the viscosity of solutions in the system N, N-dimethylformamide-phosphoric acid measured with the accuracy up to 1% demonstrate a positive deviation in the range of compositions as a whole. The degree of the existing link with complex stability suggests that the stability of hydrogen bonds in this series is declining. We can therefore draw the conclusion that the six-coordinate state in the solutions under consideration results from bond vacancies of oxygen atom and hydrogen atom linked to carbonyl group, same oxygen atom and carbon atom of methyl group, nitrogen atom and carbon atom of methyl group, or hydrogen atom of amino group, respectively. The outcomes of the unconventional approach to thermodynamic analysis and the recognized laws of the fluctuation in the energy of hydrogen bonds and the specific interactions in the liquid's acetonitrile, dimethyl sulfoxide, N, N-dimethylformamide, N methylformamide, and formamide serve as evidence for the viability of this approach and the appropriate interpretation of the nature of the interactions. Thus, the rejection of the sp_3-hybridization model made it possible to apply theoretical chemistry principles for the first time in the estimation of the energies of various types of interactions and hydrogen bonds in organic compounds.

Two HOMOs are centered on a basic oxygen molecule. In contrast to the water molecule giving the oxygen atom a lower electron density, the n-s-conjugation in $O(CH_3)_2$ contributes 1.5 eV to the HOMO destabilization. Because of this, the energy of the intermolecular interaction should be much lower than the energy of the hydrogen bond in water, which is 10.94 kJ/mol. The structure of the ether molecule in both its liquid and solid states is determined by its angular structure, which includes two oxygen atoms with lone electron pairs and terminal methyl groups. Four vacancies exist in the ether molecule; the oxygen atom fills two of them, and the carbon atoms of the terminal methyl of the alkyl groups fill the other two. The angular ether molecules form a chain of cycles where every two successive terminal methyl groups are located on perpendicular planes or a network where four specific intermolecular interactions of the same kind $-C–O \rightarrow CH_3–O–$ ($-C–O \rightarrow CH_3–O–$) are present and are primarily made up of lone $2S_2$-electron pairs of oxygen with two terminal methyl groups. The low enthalpy of vaporization also points to the absence of dimerization in the vapor. The thermodynamic calculations of the energy of certain intermolecular interactions are made easier by the molecules in vapor. The vaporization enthalpy with the symmetric alkyl groups can be used to estimate the energy of this interaction.

7.8. REVERSE DATIVE BOND

The reverse dative bond (Figure 7.2) contributes the most to the dissociation energy of the C-O bond, the specific interaction for dimethyl ether $O \rightarrow CH_3$ has a low value. According to the results of the quantum-chemical calculations, the energy of this interaction in the dipropyl ether with a normal chain (8.92 kJ/mol) is lower by 2.08 kJ/mol than the energy of the hydrogen bond in water (10.94 kJ/mol). However, the equal contribution of the reverse dative bond in the dipropyl and dibutyl ethers of the structure shows that their individual intermolecular interactions have either the same energy or a negligible difference. Therefore, the extra energy contribution from the CH_2 group into the enthalpy characteristics is indicated by the computed value of the dibutyl ether energy exceeding the energy of the interaction of the dipropyl ether in the condensed state. The interaction of hydrogen and carbon atoms of the CH_2 groups and the contribution of the energy of their specific interactions into the enthalpy characteristics of the vaporization processes of the other ethers and organic compounds with an alkyl chain exceeding three carbon atoms are explained by the absence of effect of the reverse dative bond on the terminal methyl group in the dibutyl ether. By measuring the energies of intermolecular interactions in liquid dimethyl, diethyl, and dipropyl ethers, it is possible to calculate the vaporization enthalpy of various alkyl ethers. Propyl methyl (1.60 kJ/mol) and butyl methyl (4.20 kJ/mol) ethers exhibit the difference between experimental values and those computed using the additivity method. This difference may result from two factors: first, the reverse dative bond's contribution to the methyl group ability to retain its property; and second, the deformation of the cycles in the cyclic chain structure brought on by the different lengths of the alkyl groups when two molecules are present in close proximity.

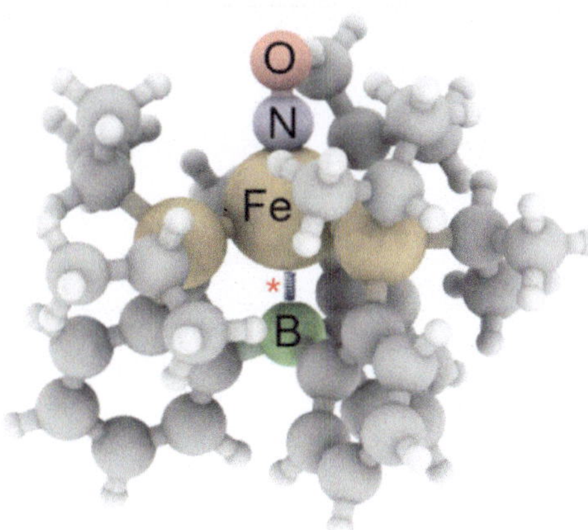

Figure 7.2. Reverse dative bond.

Source: https://pubs.acs.org/cms/10.1021/acs.inorgchem.1c02927/asset/images/medium/ic1c02927_0004.gif, Creative Commons License.

We need to know the relationship between the enthalpy properties and the quantity of carbon atoms in alkyl groups. These correlations are based on the information currently available regarding the vaporization enthalpies of symmetric and unsymmetrical ethyl and methyl alkyl ethers. Only at a large number of compounds belonging to the same homologous series could the gradually increasing influence of the interactions lead to a slight departure from the description of the experimental values of the enthalpy characteristics. This makes clear the fundamental changes in the intermolecular interaction resulting from the reverse dative bond (diminished) influence on the energy of the interaction brought on by the terminal methyl of the corresponding alkyl group in the first three members of this homologous series. The presence of two equal alkyl groups in each ether of the series results in the decrease in the influence of the reverse dative bond with the successive expansion of the number of carbon atoms in the alkyl fragment in the series dimethyl–dipropyl ethers.

The tert-butyl methyl group creates the least amount of the vaporization enthalpy when the isostructural methyl group is located at the first carbon atom (Baev, 2012). In contrast to the asymmetric isopropyl methyl ether, which has a methyl group at the oxygen, the attachment of two isostructural methyl groups significantly alters the electron density distribution, which results in diminished negative charges on both of the isostructural methyl carbon atoms. It is reasonable to anticipate that the negative charge on the carbon atoms of the isostructural methyl groups tends to grow increasing the low-stability specific bonds produced by these groups when the latter is replaced by isopropyl in the diisopropyl ether. It was proven that the polymorphic transition, melting, sublimation, and vaporization were dissociation processes and related to the dissociation processes of dimeric and more complex molecular forms in the gas phase, complex formation, and disproportionation by the steps of substitution and oxidation. The entropy of these processes was the defining factor of the structural transformations. This assertion is true for molecules that contain certain intermolecular interactions involving a pentacoordinate carbon atom as well as molecules with ionic, covalent, hydrogen, and other link types. Monomer molecules that are present in the vapor and those that form one particular sort of intermolecular interaction are not involved in the polymorphic transitions. Significant changes in the crystal structure are indicated by the enthalpy being one-fourth of the vaporization enthalpy value. Similar to what was found for the vaporization process, the reported enthalpy characteristics of the ether sublimation show their reliance on the number of carbon atoms in the alkyl chain. Since the effect of the reverse dative bond is reduced as one transition from methyl to propyl ether derivatives, it is possible to state that the variation in the enthalpy characteristics of vaporization and sublimation as a function of the number of C atoms in the alkyl group are described by the same dependencies.

In order to acquire the values within the existing precision of the experimental methods, we can thus estimate the enthalpy features of the vaporization and sublimation of a variety of ethers using interpolation and extrapolation methods. The estimation of the energies of the same interactions in crystals can be made using the calculation of the energy interactions of liquid symmetric ethers. Specific interactions in solid ethers are stabilized at 2.0–2.7 kJ/mol when compared to similar interactions in liquid ethers. The effect of the reverse dative bond in the interaction of an alkyl group containing methylene units ends at the propyl group, so in the case of dibutyl and butyl methyl ethers, it is vital to consider the contribution to the enthalpy characteristic of two or one CH_2-group, respectively, when the propyl groups contribute equally to the specific interactions. The specific interactions between butyl methyl and tert-butyl methyl ethers in both solid and liquid form are comparable to those between the propyl group and the ethyl group, respectively. These methyl substituents offer the precise interactions with equivalent energy. As a result, the energy difference between the specific interactions originating from the propyl group and the ethyl group, as well as the specific interactions of the isostructural methyl groups, is what causes the difference in the enthalpy characteristics of butyl methyl and tert-butyl methyl ethers (Tong, Benson, Wang, & Lu, 1996).

The position of the isostructural methyl group at the carbon atom of the two-carbon chains that is immediately connected to the oxygen atom is what distinguishes isopropyl methyl from tert-butyl methyl and diisopropyl ethers. Comparing the energy of specific interactions originating from isostructural methyl groups reveals that when the isopropyl in the diisopropyl ether was replaced by a methyl, the energy of this interaction was destabilized three-fold in tert-butyl methyl ether and 1.5 times in isopropyl methyl ether. As a result, the distribution of electron density in these molecules and, consequently, the distribution of charges on the isostructural methyl groups are both greatly influenced by the methyl bond to the oxygen atom. The subsequent CH_2 groups equally contribute to the enthalpy of ethers up to dodecyl methyl ether and at least to dihexyl and hexyl vinyl ethers, which is the second distinguishing trait of saturated ethers. This indicates that in a significant number of ethers in each homologous series, the specific characteristic, and the energy of the interaction of the essentially unshared $2S_2$-electron pair of the carbon atom and the positively charged hydrogen atoms remains constant. The energy of intermolecular interaction approaches its limiting value in molecules with the propyl alkyl chain. The validity of this was demonstrated in all ethers with experimentally measured and estimated interpolation and extrapolation enthalpy characteristics by the results of thermodynamic calculations of the energies of the specific intermolecular interactions that took the contribution of CH_2 groups into account. Similar to liquid and solid saturated ethers, two straight lines intersect at the point where the vaporization enthalpy of vinyl propyl ether happens to describe the dependence of the vaporization enthalpy of ethers with the smallest unsaturated alkyl on the number of carbon atoms in the saturated alkyl chain. This is because the energy of interaction emanating from the terminal methyl of the propyl group is affected by the termination of the reverse dative bond effect. These dependencies enable extrapolation and interpolation to be used to estimate the unmeasured enthalpy properties of vinyl ethyl, vinyl heptyl, and vinyl octyl ethers. The method of additive chemical thermodynamics, as in the cases before, relies on the energy contribution to the value of the vaporization enthalpy characteristic of all existing bond vacancies in the liquid and solid states. This is how the energies of specific intermolecular interactions are evaluated.

The participation of the pentacoordinate carbon atom of terminal methyl and isostructural CH_3 groups determines the nature of the specific intermolecular interaction of all crystalline and liquid ethers. As a result, there should always be a correlation between the energies of the specific interactions and enthalpy properties in different homologous series. The similarity of this correlation to straight leads to the inference that the influence of the reverse dative bond, which is seen in the ether molecules of the first three compounds in each series, is leveled in each homologous series, and that the subsequent contributions of each CH_2 group, with their increasing number in the alkyl chain, remain constant for the same reason, i.e., the invariability of the nature of the specific interactions. The oxygen in ether and cyclic saturated and unsaturated oxide molecules has two free bond vacancies, and the carbon atoms that form donor-acceptor interactions have essentially unshared $2S_2$ electron pairs. Cycles in saturated oxides provide a system that prevents the electron density from the central CH_2 group in oxetane and two similar groups in oxolane from being displaced to the oxygen atom. As the number of CH_2 groups in the molecule increases from two to four, represent the stability of the particular interactions in this series, the vaporization enthalpy increases in the series from ethylene oxide through oxetane and to 1,4-dioxane. When ethylene oxide and methyl oxirane have the same amount of free bond vacancies, comparing their structures reveals that their vaporization enthalpies (1.12 kJ/mol) are different (Moldoveanu, 2010).

7.9. INTERACTIONS

It is vital to know that the hydrogen atom that was substituted is not a part of the interaction. Because of this, the energy value in question, which is the difference between the vaporization enthalpy of ethylene oxide lacking these methyl groups and methyl oxirane with a single CH_3 group, is a sum of the energy contributed by the methyl group and the compensated energy provided by the hydrogen atom that was replaced. The number of weak specific interactions created by these groups is double for each molecule because the molecule participates in the interaction not only through its own isostructural methyl groups, but also through bonds with comparable groups of another molecule in the immediate vicinity. The difference in vaporization enthalpy between ethylene oxide and methyl oxirane should be attributed to the contribution of one of these methyl groups, and the neighboring molecule should balance out the energy of the substituted hydrogen atom interaction. Oxolane (32.90 kJ/mol) and 2-methyltetrahydrofuran (34.00 kJ/mol) exhibit similar sequence of molecular structure alterations and vaporization enthalpy values. As a result, two methyl groups from the interacting molecules contribute to the difference in the vaporization enthalpy of these compounds (1.10 kJ/mol). As a result, it is feasible to say that each methyl group participates in the intermolecular interactions in liquid saturated oxides, which has an energy value of 1.10 kJ/mol and an overall value of 2.2 kJ/mol. As a result, the total energy of stable specific intermolecular interaction in liquid oxolane is 32.90 kJ/mol, while it is 31.80 kJ/mol in 2-methyl tetrahydrofuran. The number of bond vacancies is high in the second group of oxide compounds in the dioxane series, where four of them combine to produce two oxygen atoms and four CH_2 groups. The structure of the liquid state is determined by the regular arrangement of oxygen atoms and two ethylene groups in the structure of the 1,4-dioxane molecule. Each of the four molecules in the immediate vicinity should establish eight identical-type bonds. Consequently, the breakdown of eight intermolecular interactions is where the vaporization enthalpy comes from. Since there are more bond vacancies in the 1,4-dioxane molecule than in the ethylene oxide molecule, but the number of CH_2 groups per oxygen atom and the number of $CH_2–CH_2$ bonds are calculated to be equal, it is likely that their specific intermolecular interactions have same energies.

The specific interactions are less stable in liquid 1,4-dioxane. In contrast to the intermolecular interactions present in the liquid tetrahydrofuran, the 1,4-dioxane and 1,3-dioxane molecules formed eight distinct interactions that destabilized each bond, lowering the energy by almost a factor of two. Another interesting question is how the position of the CH_2 group in the molecule affects the energy of interaction under the assumption that the number of oxygen atoms and the number of methylene groups are both identical. The thermodynamic calculations of the energies of the specific intermolecular interactions in 1,4-dioxane and 1,3-dioxane are used to demonstrate this. Each molecule of 1,4-dioxane and 1,3-dioxane bonds with four other molecules in close proximity in liquid state. The 3,5,7-trioxacyclononane molecule's oxygen and CH_2 content should cause changes in the electron densities of the oxygen and carbon atoms as well as the number of contact molecules in the immediate vicinity. The fragment of four CH_2-groups between each oxygen atom in the 3,5,7-trioxacyclononane molecule shifts the electron density to each oxygen. The cyclic structure should enhance the dative bond function; as a result, energy of the intermolecular interaction ${}_{14}O\rightarrow CH_2$ should be less stable than in liquid 1,4-dioxane and 1,3-dioxane. The 1,3,5-trioxane molecule's flat structure allows the production of net structures in both the solid and liquid states, wherein the six molecules in close proximity coordinate to allow oxygen atoms to engage their lone electron pairs with two CH_2-groups. The energy for the intermolecular interaction $O\rightarrow CH_2$ is provided by the molecule's 12 bond vacancies.

CHAPTER

8 Medicinal Organic Chemistry

Contents

8.1. INTRODUCTION

An organic molecule will dissolve either in water or in a non-aqueous lipid solvent, i.e., it won't dissolve and will appear at the interface between water and lipid solvent. A molecule is said to be hydrophilic or to have hydrophilic nature if it can completely or partially dissolve in water. An organic molecule is considered to be lipophilic or to have lipophilic property if it can completely or partially dissolve in a nonaqueous or lipid solvent. The phrases "lipophilic" or "lipid-loving" are interchangeable with the terms "hydrophobic" or "water-hating." A drug/chemical must have the ability to dissolve in both lipids and water. A drug needs to cross the GI tract lipid membrane in order to exit the digestive system (GI) (Hirtz, 1985). The chemical will subsequently enter the water-filled circulatory stream. Once in the body, the medicine can be dispersed until it reaches the target cells on or in which it is intended to function. The medicine must pass the lipid membrane to reach the aqueous intracellular fluid, or cytosol, in order to enter the cell. Drugs after passing membranes get eliminated through the kidney in the form of urine or into the GI tract in the form of biliary fluids. Solubility in lipid and water can be considered crucial for drug activity. If a molecule and its functional groups may form bond with either the molecules of the water or the lipid solvent, it can be predicted whether the chemical will dissolve in water or lipid solvent.

8.2. FORCES

Van der Waals forces, (Figure 8.1) which are electrostatic in nature, are the weakest kind of interaction. This form of attraction happens between nonpolar parts of molecules and is caused by a mutual distortion of the electron clouds that make up the covalent bonds. The induced dipole-induced dipole attraction is another name for this interaction. In addition to being weak, it also depends on temperature, being significant at low temperatures but insignificant at high temperatures. Since the attraction only lasts a short distance, the molecules must be packed tightly. Steric elements, like molecular branching, have a significant impact on van der Waals attraction. The hydrocarbon part of organic molecules like aliphatic, alkene, and aromatic systems is where this type of force is common. For each atom involved, van der Waals forces range from 0.5 to 1.0 kcal/mole. Although Van der Waals bonds exist in lipophilic solvents, they are not very significant in water. The hydrogen bond is an example of a dipole-dipole bond, which is a more powerful type of chemical bonding. The unequal sharing of a pair of electrons that make up a covalent bond causes a dipole to form. When the electronegativity of the two atoms that make up the covalent bond varies significantly, this happens. This area of the molecule becomes partially ionic, creating a permanent dipole and classifying the substance as a polar compound. The negative end of one dipole is electrostatically drawn to the positive end of the other dipole, resulting in the dipole-dipole attraction between two polar molecules.

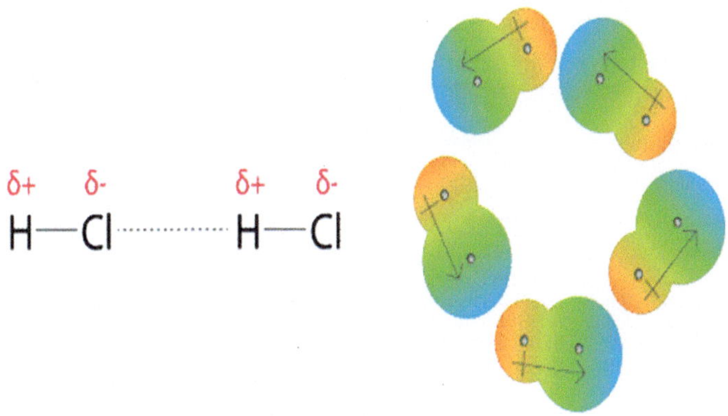

Figure 8.1. Van der Waals forces.

Source: https://chem.libretexts.org/@api/deki/files/78330/CK12_Screen-shot_9-15-1.png?revision=1&size=bestfit&width=509&height=223, Creative Commons License.

A hydrogen bond can form when an electropositive hydrogen, which is attracted to a region of high electron density, is present in at least one dipole. This can be observed when hydrogen is covalently bound to an electronegative atom like oxygen, sulfur, nitrogen, or selenium. Atoms with unshared pairs of electrons, such as amine nitrogen, ether, or alcohol oxygen, and thioether or thiol sulfurs, have higher electron densities. Although dipole-dipole bonding includes hydrogen bonding, not all dipole-dipole bonding involves hydrogen bonding. A good example of a hydrogen-bonding solvent is water, which is a key component of many drugs. The unusual high boiling point (BP) of water and the distinctive dissolving characteristics of water are both explained by the ability of water to form hydrogen bonds. Temperature and distance both impact the hydrogen bond. Each hydrogen bonding energy ranges from 1.0 to 10.0 kcal/mole. Strong attraction known as ion-dipole bonding is generally unaffected by temperature or distance. An ionic salt may occur when an organic substance having basic characteristics (amine) is put into an aqueous acidic medium ($pH < 7.0$). The ionic salt will have increased water solubility due to ion-dipole interaction. The same is true when an organic compound with acidic properties such as carboxylic acids, phenols, unsubstituted or monosubstituted sulfonamides, and unsubstituted imides is added to an aqueous basic medium (pH>7.0). In this case, the compound may form an ionic salt, which, if isolated, will have increased water solubility due to ion-dipole bonding (Rendleman, 1967).

It is crucial to know whether a drug will dissolve in water while analyzing it structurally. The quantity of hydrophilic groups in a molecule must be balanced against the presence of lipophilic groups in order to anticipate water solubility. If a molecule has a strong affinity for water, such as through hydrogen bonding or ion-dipole attraction, one will anticipate that it will dissolve in water. A molecule will dissolve in a nonaqueous or lipophilic medium if it lacks hydrophilic groups but has a lipophilic part capable of van der Waals attraction. Methane is the first system in the conventional nomenclature, followed by ethane, propane, butane, and so on. The suffix "-ane" designates the molecule as an alkane. Although there are only two ways to assemble a molecule in butane, there are several isomers that can be created, making the nomenclature difficult to use as we approach larger molecules. Consequently, a more organized nomenclature is required. The recognized nomenclature is that of the International Union of Pure and Applied Chemistry (IUPAC). The longest continuous alkane chain must be found according to IUPAC terminology. The base name is the name of this alkane chain. The chain is then assigned a number that gives the substituents the lowest possible number. The base name of the straight-chain alkane is then preceded by the number, followed by the names of each substituent.

Given its high BP, water is a medium with significant intermolecular bonding, i.e., high with respect to its molecular weight. Foreign atoms must break into this lattice in order to dissolve in or combine with water. One molecule that can do this is sodium chloride or table salt, which is water-soluble. Since an alkane cannot bond to water, it cannot penetrate the water lattice. Alkanes cannot interact with ions in an ion-dipole manner, whereas sodium chloride may. It is also impossible for water and alkane to form hydrogen or ionic bonds. With little to no van der Waals attraction between the water and the alkane, the van der Waals bonding between alkanes is relatively strong. Overall, the alkane separates off and becomes immiscible with water. In an oil or lipid solvent, alkanes will dissolve. From the perspective of solubility, the terms "lipid," "fat," or "oil" refer to a material that is either water-immiscible or water-insoluble. Alkane groups are prevalent in lipid solvents; therefore, it is not surprising that alkanes are soluble in lipid layers given the prevalence of induced dipole-induced dipole bonding. An alkane will choose to go to the lipid area if allowed to leave the aqueous region. This indicates that n-heptane will partition into Decane when it is added to a separatory funnel that also contains water. The general anesthetic alkanes and their quick dissolution into the lipid component of the brain, while at the same time having limited affinity for the aqueous blood, are examples of this migration of alkanes, which also occurs in biological systems. Chemical stability should also be listed as a quality.

8.3. TEMPERATURE

Alkenes and alkanes have similar physical characteristics. At normal temperatures, the lower members, which contain two through four carbon atoms, are gases. Alkenes with five carbon atoms or more are liquids, and as their molecular weight rises, so do their boiling temperatures. The weak intermolecular interaction is what causes the low BP. A prediction of nonaqueous versus aqueous solubility is also made possible by knowing what kind of intermolecular interaction is possible. Alkenes cannot dissolve in the aqueous layer because they cannot form hydrogen bonds and have a weak permanent dipole. In nonpolar solvents like lipid, fat, or oil layers, alkenes will dissolve. As a result, alkene's physical characteristics are similar to those of alkanes. The molecule has a reactive site due to the multiple bonds. Alkenes are susceptible to oxidation in the context of medicine, which might result in the creation of peroxide or hydroperoxide. If such a reaction happens, gaseous alkenes are the first to get oxidized. Because they are so unstable, peroxides could explode. Alkenes are extremely combustible and have the potential to explode in the presence of oxygen, especially the volatile members. Similar to the previously mentioned alkanes, the metabolism of the alkenes is uncommon. The alkene functional group should be regarded as metabolically stable for our purposes. While alkene-containing medications are often stable in the body, somebody metabolite alkene functional groups act as reaction sites. Alcohols are produced by mixing water with unsaturated fatty acids. During the production of steroids, a cytochrome P_{450} oxidase attacks the alkene functional group in squalene to produce an epoxide. Eicosatrienoate, a triene, is converted into a peroxide intermediate during the prostaglandin biosynthesis process, and alkenes are *in vivo* reduced during the formation of saturated fatty acids. As a result, we should be aware of these potential alkene functional group reactions and not be surprised if an alkene-containing medicine is metabolized.

8.4. AROMATIC HYDROCARBONS

The aromatic hydrocarbons may initially appear to be no more than cyclic alkenes; however, this is not the case. We need to keep in mind that aromatic compounds have a cloud of electrons above and below the ring rather than discrete single and double bonds. Delocalized electrons that are not as easily accessible as the electrons in alkenes are present in this cloud. The chemical processes that influence alkenes are consequently less likely to occur in aromatic systems. It is not true that the formation of peroxides, a potentially major pharmacological issue with many alkenes, is an issue with aromatic hydrocarbons. The electrophilic reaction is a common reaction in aromatic systems (Liljenberg et al., 2010). In the electrophilic reaction, the positively charged, electrophile species attack the aromatic electron-dense cloud. One key electrophilic reaction, called hydroxylation, only takes place in biological systems. Although highly crucial to drug metabolism, this reaction does not take place *in vitro*. Aromatic hydrocarbons are stable when stored. These hydrocarbons are combustible and lipophilic like other hydrocarbons. But aromatic hydrocarbons exhibit a higher tendency to connect through van der Waals attraction due to their high electron density. Aromatic rings have a key role in the binding of a drug to biological proteins.

In vivo oxidation of aromatic rings, or more precisely aromatic hydroxylation, is common. Several cytochrome P_{450} isoforms frequently catalyze this process, which could start out as an epoxidation. A few instances of this highly reactive epoxide have been isolated, but most of the time, the epoxide rearranges to produce the hydroxylation product, the phenol or alcohol. The aromatic water solubility is increased by aromatic hydroxylation. In many situations, this causes the molecule to leave the body quickly, but in a few, hydroxylation may actually make the drug more active. The role of hydroxylation of aromatic hydrocarbons and its relationship to the carcinogenic qualities of aromatic hydrocarbons has been studied in great detail. The intermediate epoxides may cause cancer. The phenols created by aromatic hydroxylation can either be removed from the body as-is or they can go through phase 2 conjugation, which results in a sulfate conjugate or a glucuronide conjugate. The halogenated hydrocarbon characteristics are somewhat different from other hydrocarbons. The extremely electronegative halide linked to the carbon gives the mono haloalkanes a persistent dipole. Intermolecular bonding is weak and dependent on the van der Waals attraction because, despite the halogen's high electron density, there are no regions that are significantly lacking in electrons. These compounds have low boiling temperatures and insufficient water solubility because only van der Waals bonding is feasible. Halogens that are covalently attached to carbon generally make their bound molecules more lipophilic. A decrease in flammability with more halogens is another characteristic of halogenated hydrocarbons. Fire extinguishers have been utilized with carbon tetrachloride. These substances are lipid-soluble and chemically inert.

When oxygen and heat are present, chloroform is transformed into the poisonous and reactive chemical phosgene. A minute amount of alcohol is added to bottles of chloroform to eliminate any phosgene that may develop. Alcohol and phosgene combine to form a harmless carbonate. It should be highlighted that only volatile short-chain polyhalogenated chemicals can undergo this oxidative reaction. In a large organic molecule, a polyhalogenated carbon can't go through oxidative dehalogenation. Because of the stability *in vivo*, there is no chemical reactivity *in vitro*. Halogenated hydrocarbons are typically difficult to break down, and this is particularly true of halogenated aromatic compounds. The halogenated general anesthetics are an exception to the stability of haloalkanes. These medicines frequently undergo oxidative dehydrohalogenation, which is catalyzed by cytochrome P_{450}. The reaction appears to be dependent on the stabilizing effect of the nearby halogen and its unshared pairs of electrons and requires at least one hydrogen bonded to the carbon holding the halogen. A halohydrin is first produced by oxidation, which then eliminates a hydrohalic acid. This reaction proceeds more quickly when two nearby trihalo or dihalo carbon moieties are present. Dehydrohalogenation intermediate is linked to nephrotoxicity as well as hepatotoxicity. The halogenated hydrocarbons have a longer biological half-life and are fat soluble, which makes it more likely that they may cause systemic toxicity. They are also not easily eliminated by the kidney. This may also explain why some halogenated hydrocarbons have the potential to cause cancer. All hydrocarbons have a key characteristic that makes them either lipophilic or hydrophobic: they are unable to form bonds with water.

8.5. INTERMOLECULAR H BONDING

Intermolecular hydrogen bonding can involve OH groups. A persistent dipole is created by the electronegativity of the oxygen and the electropositive proton (Henrici-Olivé & Olivé, 1979). The oxygen is negative in nature and the hydrogen connected to it is slightly positive. There is an uneven sharing of the pair of electrons that make up the covalent bond. Alcohol molecules have relatively high boiling temperatures compared to their hydrocarbon counterparts due to the intermolecular hydrogen bonding that is available between the alcohol molecules. The ability of the alcohol group to form hydrogen bonds with water is also significant. The alcohol functional group enhances water solubility since it may penetrate the water lattice. Each level of water solubility will vary depending on the amount of hydrocarbon present. Alcohols in the C_1–C_3 range are miscible with water in all ratios. The hydrophilic properties of the molecule diminish as the hydrocarbon chain lengthens. Even though it does not have as much of an impact as chain length, the position of the hydroxyl radical nevertheless affects water solubility. It is more likely that a hydroxyl group located in the middle of a molecule than one located at the end of a straight chain will make a substance more water-soluble. The solubility rises with the addition of a second hydroxyl group. It can be compared to a mixture of ethanol and propanol. It would be expected that 1,5-pentanediol would likewise be water-soluble. The solubility of alcohol in nonaqueous mediums would rise as the solubility of alcohol in water declines. An alcohol functional group has the power to solubilize an alkane chain with five or six carbon atoms to the level of 1% or more.

From a pharmacological perspective, the alcohol functional group is a relatively stable unit. But primary alcohol will oxidize to a carboxylic acid in the presence of oxidizing agents after passing through an intermediate aldehyde. A tertiary alcohol is resilient to mild oxidation, but the secondary alcohols can be oxidized to a ketone. Because there are so few oxidizing agents employed in pharmaceuticals, it is uncommon to meet the oxidation of alcohol *in vitro*. Although the alcohol functional group is reasonably stable *in vitro*, it is quickly broken down by a number of enzymes in the body, most notably alcohol dehydrogenase and the cytochrome P_{450} enzymes. The prevalent oxidizing enzyme is alcohol dehydrogenase, a NAD-specific enzyme. By being subjected to oxidation by oxidase enzymes, primary and secondary alcohols can both produce carboxylic acids or ketones. The most frequent phase 2 type metabolism for alcohols is the glucuronidation process. Alcohol and activated glucuronic acid, or UDP-glucuronate (UDPGA), are conjugated in this reaction to create a glucuronide. UDP-glucuronosyltransferase is the enzyme responsible for catalyzing this reaction. UGT1 and UGT2, two sets of genes with a range of isoforms within each, are responsible for producing the UGT proteins in humans. In most cases, a large increase in water solubility in the resulting glucuronide leads to a drop-in biological activity and an increase in excretion rates. UGT is mainly detected in the GI tract and liver. It should be noted that glucuronidation can sometimes lead to an increase in biological activity. The morphine-6-O-glucuronide is 650 times more potent than morphine.

8.6. SULFOTRANSFERASES

Sulfotransferases, SULT1 or SULT2, (Figure 8.2) enzymes present in the liver, gut, kidney, and brain tissue, catalyze this process (Coughtrie, 2016). The solubility of water increases in the sulfate conjugates as well. The higher water solubility is due to the hydrogen and ion-dipole bonds. The ability of the permanent dipole to form intermolecular hydrogen bonds leads to high boiling temperatures and water solubility. The only way cyclohexanol differs from phenol is by not having an aromatic ring. The difference between cyclohexanol and phenol in terms of BP and solubility may appear significant, and it should be given that phenols have a key feature that alcohols lack acidity. Let's look at some other parameters that influence solubility before talking about the phenols' acidity. The phenol's water solubility decreases as its lipophilic nature rises. The water solubility is significantly decreased by the addition of a methyl (cresol) or halogen (chlorophenol). Similar to the diols, adding a second hydroxyl increases solubility in water.

With the addition of alkyl to this molecule, catechol solubility will drastically decline. An acid is any substance that has the capacity to release a proton. Because of this property, phenol can be regarded as an acid. Because of their acidity, phenols can form stronger ion-dipole bonds with water in addition to hydrogen bonds since they exist both as neutral molecules and, to some extent, as ions. The presence of both dipole-dipole bonding and ion-dipole interaction is related to the prediction of a higher BP and increased water solubility. The replacement of the aromatic ring affects the acidity of phenols. While meta or para substitutions to the phenol produce known acidities, ortho substitution to the phenol has unpredictable effects on acidity. Acidity is reduced by substitution with a group that can donate electrons to the aromatic ring. When the substitution is para or in the direct conjugation, the effect is the most noticeable. Acidity rises when an electron-withdrawing group is added to the aromatic ring. Again, the para replacement has the most noticeable effect. In both situations, the ability or incapacity of the substituent to stabilize the phenolate state determines how the substituent affects acidity. The chemical reactivity of phenols is an important characteristic. Being a weak acid, phenol won't react with sodium bicarbonate, a weak basic, but will instead react with strong bases like sodium hydroxide or potassium hydroxide to produce the corresponding phenolate salts.

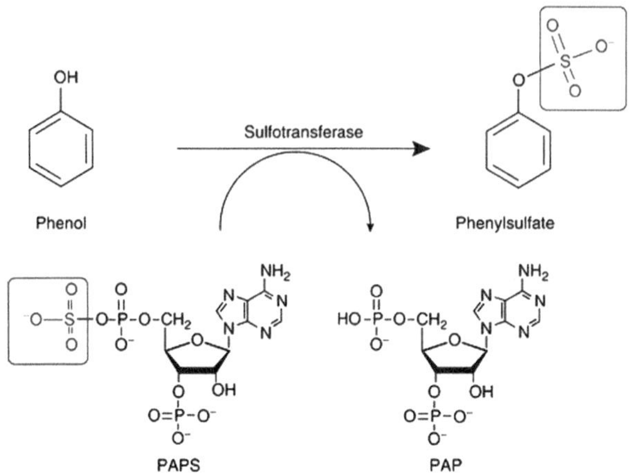

Figure 8.2. Sulfotransferases.

Source: https://media.springernature.com/lw685/springer-static/image/prt%3A978-3-642-16483-5%2F19/MediaObjects/978-3-642-16483-5_19_Part_Fig1-5568_HTML.gif, Creative Commons License.

Since the produced phenolates are ions and will dissolve in water due to the stronger ion-dipole interaction, salt production is a crucial process. The simple phenols (phenol, cresol, and chlorophenol) are very soluble in water when they are salts. The phenols' solubility in water will be significantly increased by sodium and potassium salts. Due to the salt's inability to dissociate in water, heavy metal salts of phenols will actually become less water-soluble. The solubility of phenols in water is always increased by their salts, and for the majority of phenols with therapeutic value, the salts provide sufficient solubility for the medicine to dissolve in water at the concentration required for biologic activity. The solubility of the phenolate salts will, however, diminish as the number of lipophilic components linked to the aromatic ring increases. Despite the fact that salt production is a type of chemical reaction, it is not a chemical instability. The water-soluble salt will undergo acid treatment to stop this reaction and regenerate phenol.

8.7. PHENOL

The simple oxidation of phenols is a second important chemical process. Quinones are intensely colored and formed when phenols are oxidized. Due to the creation of p-quinone or o-quinone, a clear solution of phenol that is kept in contact with air or light quickly takes on a yellow color. This reaction happens more easily with polyphenolic chemicals and phenolic salts. When kept in closed, amber containers or with the addition of antioxidants, phenols, and their salts must be shielded from oxygen and light. Alcohol and phenol metabolisms are very similar. To create a phenolic chemical, the phenol may be oxidized. The new OH group will often be orthogonal to the original hydroxyl group. The cytochrome P_{450} family of enzymes often catalyzes hydroxylation processes. The two processes that frequently convert phenols into products are conjugation with glucuronic acid to create glucuronides and sulfonation to produce sulfate conjugates. Both conjugation processes produce metabolites with higher water solubilities than the phenol itself. Methylation is a different type of metabolism that is frequent with endogenous substrates and is exhibited to a lesser amount in xenobiotics. S-adenosylmethionine (SAM) is a cofactor required for this process. Methyl ether has lower water solubility. The OH groups in diethyl ether, which is produced by combining two moles of ethanol, have been lost. The absence of the OH group prevents the formation of hydrogen bonds between molecules, leaving only weak van der Waals attraction and a low BP.

Hydrogen will form a connection with oxygen electrons. As a result, ethers exhibit partial water solubility; but, as the hydrocarbon content rises, the water solubility rapidly declines. The onset and duration of biological activity in the context of general anesthetics are significantly impacted by water solubility. The ethers are stable, nonreactive substances. When in contact with ambient oxygen, liquid ethers produce peroxides. Although not in significant amounts, the peroxide generated can be highly irritating to the mucous membranes and, if concentrated, may explode. As a result, care should be ensured when handling ethers to reduce oxygen interaction. In order to absorb oxygen and avoid this instability, an antioxidant, such as copper metal, is frequently added. Ether metabolism is generally simple. The metabolic dealkylation reaction is important. When this does occur, the lost alkyl group is typically a methyl or ethyl group. In the dealkylation of an ether, a phenol is formed, which is subsequently metabolized via glucuronide or sulfate conjugation and catalyzed by cytochrome P_{450} enzymes. If the alkyl radical is ethyl, the alkyl group is lost as an aldehyde, formaldehyde, or acetaldehyde. The alkyl or arylthio group, also known as the alkyl or aryl mercaptans, is another name for the thioethers.

Depending on the type of thio substituents, thioethers can go through one of two metabolic processes. The peculiar metabolic oxidation response that thioethers may go through is important. Since sulfur can exist in a number of oxidation states, including 2, 4, and 6, the oxidation reaction may be important. The thioether moiety could either be oxidized to sulfoxide or sulfone. Flavin mono-oxygenase is the catalyst for this process. Short-chain alkyl thioethers may be subject to dealkylation, which results in the production of an aldehyde and a thiol, in a manner similar to that experienced by short-chain alkyl ethers. The enzymes of the cytochrome family catalyze this process. It is vital to know that both aldehydes and ketones have polar carbonyl groups, which makes the compounds polar. The electron cloud that makes up the carbon-oxygen double bond is skewed toward oxygen because oxygen has a higher electronegative charge than carbon. The enol form may coexist in equilibrium with ketones and, to a lesser extent, aldehydes. Aldehydes and ketones have higher boiling temperatures than nonpolar compounds of equivalent molecular weight due to the polar nature of the carbonyl. Aldehydes and ketones can bond with water and can partially dissolve in it because of the high electron density of the oxygen atom. Although stronger, the hydrogen bonding is comparable to that proposed for ethers. Remember that the influence of the polar carbonyl group on total solubility will diminish as the proportion of nonpolar hydrocarbons grows.

8.8. ALDEHYDES

Aldehydes are oxidized quickly because they are one oxidation state away from the stable carboxylic acid structure (Jacob, Giles, Giles, & Sies, 2003). Because oxidation occurs with many liquids, substances containing aldehydes must be shielded from ozone. The low-molecular-weight aldehydes can also be polymerized to create straight-chain polymers or cyclic trimers, a substance with three aldehyde units. Although the trimers are stable to oxygen, heating them will allow the aldehyde to regenerate. In some conditions, it is preferable to use this process to protect the aldehyde. Some drug metabolites exist as one of the products of the hemiacetals, hemiketals, and acetals or ketals. A hemiacetal is produced when an aldehyde and alcohol react in an acidic environment. Regardless of pH, the hemiacetal is unstable in aqueous conditions. An acetal can be produced by adding a second alcohol to the hemiacetal. Acetals are stable in water under neutral or basic pH conditions, but they become unstable in water under acidic conditions, converting back to the original aldehyde. A ketone can also change into a hemiketal and then a ketal when reacting with alcohol in an acidic environment. Hemiketals and ketals share the same stability characteristics as acetals. The glucuronides created when glucuronic acid is combined with alcohols or phenols are examples of acetals, although the sugar glucuronic acid itself is an example of a hemiacetal. The hemiacetal in these situations is an intramolecular hemiacetal. The acetal functional group is present in a class of drugs known as aminoglycoside antibiotics and is also present in polymers of aldehydes.

8.9. KETONES

Ketones and aldehydes have a variety of *in vivo* metabolic pathways. Xanthine oxidase, aldehyde oxidase, and NAD-specific aldehyde dehydrogenase rapidly oxidize aldehydes in general, producing carboxylic acid. These oxidizing enzymes are not linked to cytochrome P_{450}. Although an aromatic alkyl ketone may undergo an oxidation reaction to produce an aromatic acid, ketones are generally resistant to oxidation in living organisms. Aldehydes and ketones may be impacted by a metabolic process called reduction. Aldehydes appear to undergo reduction rarely, although many ketones, particularly, unsaturated ketones, go through reduction to a secondary alcohol. This reaction frequently produces only one isomer, which is stereoselective. One of the most prevalent functional groups in pharmaceuticals is the amine functional group, which serves two purposes in medicine. One function is to solubilize the medication, either as a free base or as a salt of the amine that is water soluble. The amine's second function is to serve as a binding site, keeping the medication bound to a particular location in the body where it can create biological activity. The former function helps to shape an essential amine physical characteristic. Despite being polar chemicals, amines may have low boiling temperatures and poor water solubility. One explanation for this is the absence of an electropositive group linked to nitrogen in the tertiary amine. Although there is an electropositive hydrogen bonded to nitrogen in the primary and secondary amines, the dipole formation is rare because nitrogen is not as electronegative as oxygen. All of this indicates that the intermolecular hydrogen bonding in primary, secondary, and tertiary amines is negligibly low.

It is necessary to consider a distinct aspect while determining water solubility. Due to the amine's unshared pair of electrons, the area around the nitrogen has a high electron density. Due to hydrogen bonding between hydrogen and electron-dense nitrogen, this high electron density encourages water solubility. This also happens with low-molecular-weight ethers;

however, it happens more frequently with basic amines. More soluble than secondary amines, which are more soluble than tertiary amines, are typically primary amines. The capacity to create salts of the amines is a very significant feature. A base is defined by Bronsted as a substance that can accept a proton from an acid. Amines have a pair of unshared electrons that are essentially accessible for sharing. A base's relative capacity to give its unshared pair of electrons determines how strong it is. The measure of the basicity of an amine is acquired by taking into account the acidity of the conjugate acid created by protonating the amine in order to quantify the strength of a base. The ammonium ion produced by this protonation has a p_{Ka} value. So, a strong base binds onto the proton, exists as the ammonium ion, has a low dissociation constant (Ka), and consequently has a high p_{Ka}. On the other hand, a weak base has a lower p_{Ka} because it generates an unstable ammonium ion that dissociates quickly and has a large dissociation constant (Ka), and thus readily donates its electrons.

Through their effects on electron availability, two things have an impact on basicity. One is steric, whereas the other is electronic. In the case of the former, electrons are forced into the basic nitrogen if electron-donating groups are connected to it. The electron pair on the nitrogen will be pushed out from the nitrogen because a negative charge repels a negative, making the pair more easily available for donating. The unshared pair of electrons will be drawn to the nitrogen atom and become less readily available for donating if, on the other hand, electron-withdrawing or electron-attracting groups are connected to the nitrogen. As a result, a weaker base is produced. Alkyl is a good example of an electron-donor, whereas aryl or phenyl is a good example of an electron-withdrawing group. Theoretically, primary alkyl amines with a single alkyl group connected to the nitrogen should be more basic than secondary alkyl amines with two electron-releasing groups attached to the nitrogen. Large alkyl groups can prevent hydronium ions, a source of a proton, from approaching the unshared pair of electrons. While it has little to no impact on primary and secondary amines, the steric effect becomes significant for tertiary amines. Large alkyl groups associated with amines vibrate back and forth, preventing water from approaching. As a result, salt formation does not take place as frequently as it would if these complications did not exist. Secondary amines are frequently found to be more basic than tertiary amines, and tertiary amines to be more basic than primary amines when it comes to alkyl amines.

8.10. ALKYL AMINES

Alkyl amines are substantially more basic than aromatic amines. The aromatic ring functions as an electron sink due to its delocalized cloud of electrons. Thus, the aromatic ring serves as an electron-withdrawing group. The unshared pair of electrons is resonance stabilized. The ability of the molecule to donate electrons is diminished as a result of the electron density dispersing over a larger surface area, which lowers basicity. An additional aryl or alkyl group substitution on the nitrogen of aniline predictably alters the basicity, with the aryl group reducing basicity to a neutral molecule and the alkyl group increasing basicity. The basicity is impacted by the replacement on the aromatic ring. Basicity is predictable when meta or para is substituted for the amine, while basicity is uncertain when ortho is substituted. Reduced basicity is caused by an electron-withdrawing group linked to the aromatic ring in the meta or para position. Basicity typically rises above that of aniline when electron-donating groups are in the meta or para position (higher p_{Ka}). The group basicity increases when it is in the para position and decreases when it is in the meta position. A phenol's p_{Ka} is elevated by an electron-donating group added to the meta or para position, making the phenol less acidic or more basic. Predictability fails with ortho-substituted anilines due to intramolecular interactions.

This is a crucial reaction because there is a good chance that the salts that are formed will be water-soluble if they dissociate in water. If the medicine contains a basic amine, it can be changed into a salt, which is then utilized to make aqueous solutions of the drug. Acids such as hydrochloric, sulfuric, phosphoric, tartaric, succinic, citric, and maleic are widely employed to prepare salts. Because it only has one proton and hence reacts with one base molecule, hydrochloric acid is a monobasic acid. The remaining substances are tribasic acids and dibasic acids. Some of them are sulfuric, tartaric, succinic, and maleic acids, citric and phosphoric acid. The aqueous solution of the amine salt will have pH that will vary based on the acid used. If a strong mineral acid is employed to produce salt, the pH will be acidic; otherwise, it will be weakly acidic or neutral. Since the acid transforms the amine into a salt that is water soluble, it makes sense to predict that adding a base to the salt would release the free amine, which may then precipitate. When drugs are combined, this chemical incompatibility could be very significant. The body can handle amines through a variety of metabolic pathways. Dealkylation is a typical reaction that secondary and tertiary amines go through. The alkyl group is lost as an aldehyde or ketone during the dealkylation reaction, and the amine is changed from a tertiary amine to a secondary amine and then to a primary amine. The amine is typically substituted by tiny alkyl groups like a methyl, ethyl, or propyl group to produce this reaction. Imipramine, which is converted to desipramine, is an example of a medication that undergoes a dealkylation process.

Deamination, a type of dealkylation process, can also occur with primary alkyl amines. This process results in the production of pyridoxamine, which may be catalyzed by pyridoxal 5-phosphate. A carbon bound to the nitrogen must be replaced with at least one hydrogen for this reaction to take place. The monoamine oxidase (MAO) and diamine oxidase enzymes frequently catalyze deamination processes (Tipton, Boyce, O'Sullivan, Davey, & Healy, 2004). The deamination of norepinephrine is done via the MAO-catalyzed process. MAO is not a cytochrome P_{450} enzyme, but under some conditions, primary amines may go through a dealkylation that is catalyzed by CYP_{450}. Some secondary acyclic and cyclic amines as well as tertiary acyclic and cyclic amines can also go through oxidation processes that are mediated by microsomal flavin-containing mono-oxygenase and CYP_{450} isoforms. Tertiary amines can be N-oxidized to produce an N-oxide. The methylation process is a minor metabolic pathway that is available to amines. The manufacture of epinephrine from norepinephrine, which is carried out by a number of methyltransferases, including the enzyme phenyl ethanolamine-N-methyltransferase (PNMT) and necessitates the cofactor SAM, is an example of the methylation reaction. Methylation actually results in an increase in lipophilicity. The more frequent phase 2 conjugation processes are very significant for the metabolism of primary, secondary, and a few tertiary amines. Amines and glucuronic acid can be combined to create glucuronides.

8.11. GLUCURONIDES

Glucuronide is an acetal mixed functional group. Sulfate conjugates are produced when sulfuric acid and amines are combined. The solubility in water increases for both glucuronide and sulfate conjugates. A molecule that typically exhibits a decrease in water solubility can be produced when acetyl CoA is used to acetylate amines, particularly primary aryl amines. A few tertiary amines have been observed to undergo a less frequent phase 2 glucuronidation. The resulting quaternary amine is extremely unstable and has improved water solubility. Because of this, these conjugates frequently aren't isolated; instead, the parent drug alone is found in the urine or stools. Because they are salts of a mixed acetal, they exhibit instability. Naturally occurring carboxylic acids, often known as fatty acids, are crucial to human nutrition. The fatty acid group includes stearic, palmitic, myristic, and lauric acids. It should be noted that these acids share a common biosynthetic process in which two carbon acetate units are joined together to generate naturally occurring fatty acids, which explains why they all have an even number of carbons. The fact that all of the aforementioned acids are saturated fatty acids shows that no double bonds exist and that all carbons aside from the carbon in the carboxylic acid group are completely saturated with hydrogen. The most prevalent saturated fatty acids are palmitic and stearic acids.

8.12. FATTY ACIDS

The monounsaturated fatty acids make up a second class of fatty acids. Oleic acid and palmitoleic acid are two of the most prevalent members of this class of compounds (Bakke et al., 2012). The double bond of monounsaturated fatty acids is typically found at position nine and is "cis" in stereochemistry, which is a characteristic of these fatty acids (Z-configuration). The stereochemistry of naturally occurring fatty acids is "cis," but the existence of a "trans" double bond (E-configuration) is artificial and results from food processing. The polyunsaturated fatty acids make up a third class of fatty acids. Polyunsaturated implies two or more double bonds by definition. Linolenic, linoleic, and arachidonic acids, also referred to as omega fatty acids, are the prevalent polyunsaturated fatty acids. A double bond can be located in either three or six carbons from the terminal methyl group. Since linoleic and linolenic acids are necessary for the formation of cell membranes as well as other processes, they are referred to as "essential fatty acids." The manufacture of the hormones prostaglandin, prostacyclin, and thromboxane requires arachidonic acid. Recent research has shown that a number of dietary fish oils can significantly lower the risk of coronary heart disease. Examples of fish oils include DHA and EPA, which is a precursor to prostaglandin-3. The carbonyl and hydroxyl groups that make up the carboxylic acid functional group are both polar, hydrogen-bonding groups on their own. Another carboxyl hydrogen of the OH can form a hydrogen bond with either of the oxygen groups. Because carboxylic acids are more acidic than alcohols or phenols and because there are more sites for hydrogen bonding, there is more hydrogen bonding in the case of carboxylic acids than there is in the case of alcohols or phenols. Given this explanation, one could infer that liquids and solids containing carboxylic acids have high boiling points (BPs). It is logical to assume that the carboxyl group can H-link to water, leading to water solubility, if the carboxyl can firmly H-bond to itself. The boiling temperatures of a few of the carboxylic acids show the impact of the strong intermolecular hydrogen bonding, and the solubility of the carboxylic acids in water shows the strong hydrogen bonding to water. The water solubility substantially reduces as the length of the lipophilic hydrocarbon chain increases.

8.13. SOLUBILITY

Water solubility increases in salts made from carboxylic acids and sodium, potassium, or ammonium hydroxide. Heavy metal salts have a tendency to be insoluble. The heavy metal salts of carboxylic acids, such as calcium, magnesium, zinc, and aluminum, are an example of such insoluble salts. An alkaline pH is typical when salts of carboxylic acids dissolve in water. Sodium and potassium salts typically have a pH that is relatively high. The carboxylic acid-base process can be reversed in this solution, just like with other salts, to regenerate the carboxylic acid. Precipitation could occur because the free acid is less soluble than the salt. When working with water-soluble carboxylate salts, one should be aware of this chemical incompatibility. The functional groups amines and carboxylic acids are frequently present in medications or drugs. Both of these groups can produce salts that, if capable of dissociation, can improve water solubility and have an influence on solubility. A polar alcohol and polar acid are joined to create esters, which result in a less polar liquid. When two alcohols are combined to form ethers, the boiling temperatures, and polarity both drop. Similar to others, the ester loses the two hydroxyl groups required for intermolecular hydrogen bonding. This results in a reduction in water solubility as well as the loss of intermolecular hydrogen bonding. Acetic acid has a BP of 118°C, which is higher than the BPs of several acetate esters. Acid and alcohol OH groups as well as the OH group in water form hydrogen bonds that contribute to the acids and alcohols' water solubility. With esters, the hydrogen (H_2O) and the ester carbonyl's electron-dense oxygen form a hydrogen bond. The ester has less capacity to form hydrogen bonds with water, which reduces its water solubility. Esters are not very soluble in water, but they are quite soluble in alcohol.

The ease with which most esters can hydrolyze back to the alcohol and the carboxylic acid is a crucial chemical property. Esters are particularly susceptible to hydrolysis by bases, but they can also hydrolyze in the presence of acid and water. This means that esters must be shielded against excessively alkaline conditions because they are unstable in the presence of basic media *in vitro*. The human hydrolyzing enzymes work through a mechanism that is catalyzed by bases. Therefore, it is not surprising that esters are unstable in the body and change into free acid and alcohol. Even though carboxylic acids constitute the active substance, they are frequently synthesized as esters. It is understood that the acid will undergo metabolic regeneration. The intramolecular cyclization of an amine and a carboxylic acid yields amide. The amide that results is called lactam. Similar to lactones, lactams serve as the building block for a number of pharmacological classes. The β-lactam antibiotics, which are present in penicillin and cephalosporin, are the most well-known lactam drugs. The lactams have physical-chemical and metabolic properties with open-chain amides, while a high level of reactivity is observed with tiny ring molecules like β-lactams.

8.14. CHARACTERISTICS

The physical characteristics of the amides are different. Amides are formed by mixing a polar carboxylic acid with a weak polar primary, secondary, or ammonia amine to create mono-, di-, or unsubstituted amides, just like esters. The high boiling temperatures and water solubility of the resultant amides show that they still have a lot of polarity. These characteristics are very dissimilar from esters. The BP lowers with any series as the nitrogen substitution increases. The N unshared pair of electrons has been distributed among nitrogen, carbon, and oxygen instead of remaining on the N proper. Due to the high electron density of oxygen bonding to the hydrogens or to the nitrogen, or the hydrogen bonding to the oxygen, the unsubstituted and monosubstituted amide should exhibit strong intermolecular bonding since the amount and strength of intermolecular bonding determines boiling temperatures. Since both hydrogens have been replaced on the nitrogen in the case of the disubstituted amides, intermolecular hydrogen bonding is not possible. Dipole-dipole bonding is still possible for disubstituted amides, but not via hydrogen bonding. The BP decreases as a result. Since water may provide the hydrogen, only a polar substance is needed for water solubility. Both substituted and unsubstituted amides exhibit strong water solubility and can form hydrogen bonds with water via the hydrogen in water. The amide lipophilic character grows along with its hydrocarbon content while its water solubility diminishes.

The higher stability of the amide is a chemical trait that sets it apart from esters. Amides are moderately resistant to the enzymatic, acidic, and basic environments. The resonance forms of the amide and their overlapping electron clouds may be the reason behind this stability. In order to develop drugs with sustained activity, it is advantageous to take use of the amide based increased stability over the ester. Amidase enzymes can catalyze metabolic hydrolysis even though amides are generally resistant to acid, base, and enzymes. It is crucial to know that amides are more stable *in vivo* than esters are. Although the amides were made using basic amines, because of their essentially neutral functional groups, acid salts cannot be generated. It is necessary for the unshared pair of electrons to be available for donation, which is not the case in amides. While the urea's characteristics are comparable to those of the amide, the carbonate's physical and chemical characteristics are similar to those of the ester. The physical characteristics of the carbamate reflect the combined impact of the two elements. Carbonates and carbamates are both unstable to acid and basic conditions due to the fact that their chemical reactions are more similar to those of an ester. Because they share characteristics with amides, such as polarity, the ureas are comparatively inert solids.

The aryl sulfonamides often have very low water solubility and high melting points (MPs) as solids. The chemical stability of aryl sulfonamides is comparable to that of carboxylic acid amides. Benzene sulfonamides are resistant to hydrolysis by acids, bases, and enzymes. The aryl sulfonamides are weak acids, as opposed to the carboxylic acid amides, which are regarded as neutral substances. The SO_2 moiety's capacity to stabilize the nitrogen anion via resonance accounts for the acidic character. Aryl sulfonamides will react with a strong base, like sodium or potassium hydroxide, to produce salt. Ion-dipole bonding causes the sodium or potassium salt of a sulfonamide to easily dissociate in water, producing highly water-soluble compounds. Following the dissolution of a sodium or potassium salt of a sulfonamide, the pH of the aqueous medium has a tendency to be alkaline. Since an acid-base reaction may take place and typically result in the precipitation of the aryl sulfonamide, the water-soluble salts of aryl sulfonamides are incompatible with acidic media. A class of drugs known as organic nitrates is included among the cardiac agents. Nitroglycerine is one of the more well-known drugs in this group and is actually an ester of nitric acid and triol glycerin. Because nitric acid and free alcohol can be formed through hydrolysis, the nitrate ester structure accounts for potential instability. This reaction is a potential metabolic product and can also take place *in vivo*. The nitrite functional group, although uncommon, is present in the substance amyl nitrite and is made up of an ester of nitrous acid and amyl alcohol. The release of NO seems to be the mechanism by which nitrates, and amyl nitrite cause these reactions.

8.15. ISOMERS

Two geometrical isomers may develop from the reaction of a ketone or aldehyde with hydroxylamine. The Z designation (or syn configuration) is given when the highest priority substituent and the OH group are on the same side of the double bond and the E designation is given when they are on the other side of the double bond (or anti-configuration) (Masamune, Kaiho, & Garvey, 1982). The oxime that results is typically regarded as neutral. Aldehydes or ketones react with hydrazine to form hydrazone, which is similar to oxime. In contrast to this functionality, which is known as an imine, the development of a carbon-nitrogen double bond is known as a Schiff base. When a primary amine interacts with a ketone or aldehyde in a biological system, Schiff base results. The resultant hydrazone will hydrolyze back into hydrazine and the ketone or aldehyde since it is unstable to aqueous acid. In the hydrazone, the imine nitrogen and the –NH_2 nitrogen are both basic. Hydrazones can be geometric isomers, just like oximes. Hydrazine is produced when the hydrazone is reduced. Hydrazine exhibits basic characteristics and is stable for hydrolysis by bases or acids. Acetylation of one of the nitrogen results in a hydrazide, in which the –NH_2 nitrogen maintains its basic characteristics while the –NH nitrogen behaves like a neutral amide.

8.16. EPOXIDES

Epoxides are ethers, but due to their three-membered ring, they possess peculiar characteristics. The average binding angle between the three atoms is 60°, which is significantly smaller than the typical tetrahedral bond angle of 109.5°. Thus, in the presence of either acid or base catalysts, this highly stretched ring easily opens. These interactions are significant since epoxide-containing medicines are similarly highly reactive *in vitro* and *in vivo*. In the presence of an acid, these drugs will react with a nucleophile (N:) or with a base that serves as a nucleophile to produce open-chain compounds. When epoxide-containing drugs are given to a patient, the epoxide is likely to interact with biopolymers (proteins) and have negative effects on the cell. These drugs may be used in cancer chemotherapy; however, they are frequently found to be quite hazardous. The strain energy contained in the β-lactam is what gives this molecule its distinctive trait, which is its reaction in hydrolysis by aqueous acid, aqueous base, and enzymatic conditions. The presence of the sp_2 carbonyl carbon increases this strain in addition to the four-membered ring distortion of the typical bond angles for carbon. A significant problem of many penicillins, cephalosporins, and other β-lactam containing antibiotics is the hydrolysis of the β-lactam. The five-membered ring compounds like furan and tetrahydrofuran are widely used.

There are many drugs that contain furan, but there are hardly any medications that contain tetrahydrofuran in its pure form. The ribose-containing anticancer and antiviral drugs exhibit what looks to be tetrahydrofuran; however, these sugars are actually mixed with acetal structures. Furan has characteristics similar to benzene than an ether while having an ether-like appearance. An aromatic ring is a furan. Aromatic rings are flat molecules with $4N_2$ electrons. When considering the electrons or sp_2 electrons in furan, it becomes clear that there are two pairs of sp_2 electrons on the oxygen as well as four electrons in each of the two double bonds. There is a cloud of six electrons above and below the plane of the ring because one pair of oxygen's electrons is in the same plane as the four electrons of the two double bonds. Tetrahydrofuran, however, differs greatly from furan in terms of its characteristics. Tetrahydrofuran is only a cyclic ether, but in contrast to its closest open-chain molecule, diethyl ether, it does not exhibit partial water solubility; instead, it is very soluble in water. This is because the unshared pairs of electrons on the ether oxygen and the hydrogen in water have a strong bond. Tetrahydrofuran, like the ethers, must also be shielded against atmospheric oxygen because it is easily oxidized in the presence of air to produce peroxides. Pyrrole and pyrrolidine, as well as polymers containing these compounds, undergo metabolism. Pyrrole would likely undergo aromatic hydroxylation, but pyrrolidine would be subject to sulfuric or glucuronic acid conjugation.

The correct identification of the class of compounds to which the five-membered sulfur-containing heterocycles belong is essential for understanding their properties. Since thiophene has an aromatic ring, it is more stable than tetrahydrothiophene. The sulfur-containing molecules are less water-soluble than the oxygen analogs, and the thioethers are more stable than oxygen ethers. In general, the presence of sulfur replaces oxygen and significantly reduces its hydrophilic properties while significantly increasing its lipophilic properties. Thiophene is expected to undergo aromatic hydroxylation, which can take place at any location that is substituted with hydrogen. Sulfur oxidation for reduced thiophenes, such as tetrahydrothiophene, is expected to take place. Oxazole and isoxazole are both aromatic substances. Two electrons from each double bond and two more electrons from the oxygen atom make up the aromatic cloud. Both of these compounds are basic because nitrogen retains its unshared pair of electrons, although they should be understood as weak bases. With an acid, such as hydrochloric acid, both compounds can be transformed into salts. Oxazoles and isoxazoles would be expected to be water soluble as their hydrochloric salts. Aromatic compounds with one basic and one neutral nitrogen include pyrazole and imidazole. The four electrons and the unshared pair of electrons on the NH nitrogen give rise to the aromatic character. The aromatic nuclei known as thiadiazols, triazoles, and tetrazoles often include two (triazole) or three (tetrazole) basic nitrogen in the ring. Although the triazoles are weak bases, it should be clear which nitrogen retains the electrons needed for basicity if one account for the electrons required for aromaticity.

CHAPTER

9 Enzymes

Contents

9.1. INTRODUCTION

Enzymes are organic proteins that catalyze chemical reactions and RNA can also catalyze some of the reactions in the living systems. Urease, which Sumner crystallized in 1926 and demonstrated to catalyze the hydrolysis of urea to CO_2 and NH_3, was the first enzyme identified as a protein (Balasubramanian & Ponnuraj, 2010). However, it was another 70 years until Andrew Karplus was able to determine its structure (*Klebsiella aerogenes*) (Figure 9.1).

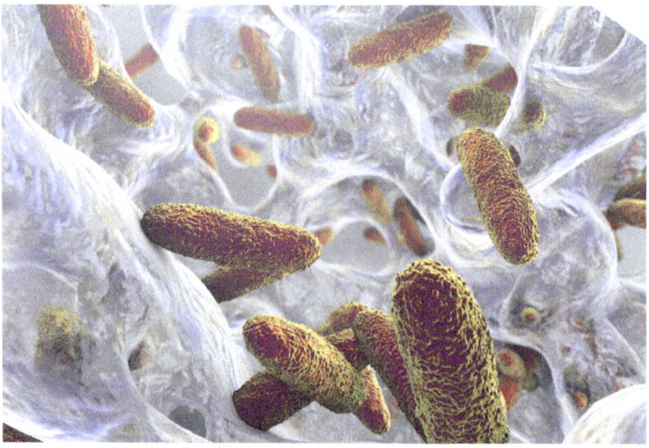

Figure 9.1. Klebsiella aerogenes.

Source: https://www.infectiousdiseaseadvisor.com/wp-content/uploads/sites/16/2019/01/bacteriaklebsiellash58654766_1362008.jpg, Creative Commons License.

One of the few Ni-containing enzymes now known is urease. Hundreds of enzymes had been found by the 1950s, and many of them had been homogeneously purified and crystallized. Ribonuclease A, an enzyme with only 124 amino acids, was sequenced for the first time in 1960 by Hirs, Moore, and Stein. Stanford Moore and William H. Stein shared the 1972 Nobel Prize in Chemistry for their development of the technology for protein sequencing that was used to establish the sequence of ribonuclease A. Despite having molecular weights between a few 1,000 and a few million, enzymes can catalyze changes in molecules as small as carbon dioxide or nitrogen. In general, enzymes work by increasing the energy of the ground state while decreasing the energies of intermediates and transition states. Similar to a chemical reaction, an enzyme-catalyzed reaction has a transition state that is high-energy and has a lifetime of approximately 10^{13} seconds, or one bond vibration.

9.2. ENZYME-SUBSTRATE COMPLEX

The development of an enzyme-substrate (or E-S) complex (Figure 9.2) in the minute cavity known as the active site, which is where the catalysis occurs, always serves as the catalyst for an enzyme-catalyzed reaction. The idea of an enzyme-substrate complex was first put forth independently by Brown and Henri in 1902; this concept expands on Fischer's lock-and-key hypothesis from 1894, which claimed that an enzyme serves as the lock that the substrate (the key) fits into. The great degree of specificity of enzymes would be explained by this interaction between the enzyme and substrate, however the lock-and-key theory does not explain all reaction events. For instance, molecules with structures similar to those of the substrate but with less bulky substituents frequently fail to be substrates while having the right dimensions to fit inside the enzyme. It has been discovered that some substances with bulkier substituents attach to the enzyme more firmly than the substrate. One would anticipate that a bulkier compound would not fit into the lock if the lock-and-key theory were to be true.

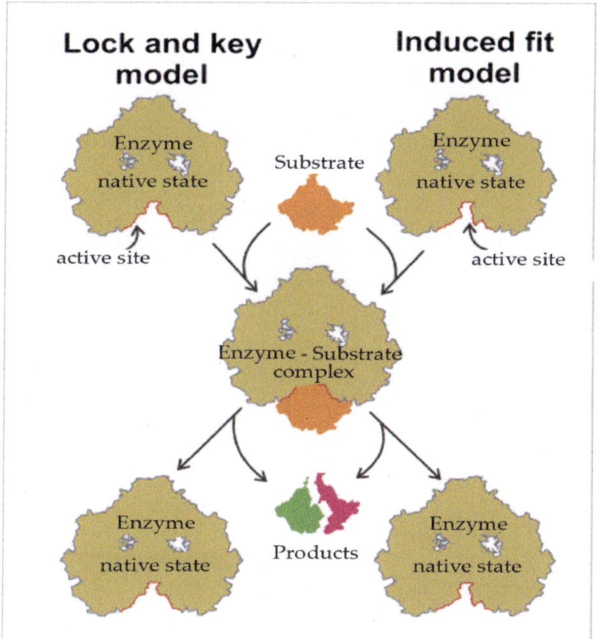

Figure 9.2. Enzyme-substrate (or E-S) complex.

Source: https://ars.els-cdn.com/content/image/3-s2.0-B9780323899291000251-f34-01-9780323899291.jpg, Creative Commons License.

When two substrates are being used in one reaction, some enzymes that catalyze the reaction only bind one substrate after the other has already been attached to the enzyme. Given these problems, Koshland put forth the induced-fit hypothesis in 1958. According to this theory, as soon as a substrate starts to bind to an enzyme, various groups on the substrate start interacting with specific enzyme functional groups, and this mutual interaction causes the enzyme to change conformation. By destabilizing the enzyme and/or causing appropriate alignment of the groups involved in catalysis, this changes the enzyme from a low catalytic form to a high catalytic form.

The conformational alteration can act as the foundation for substrate specificity. Although they may bind to the enzyme, substances that are similar to the substrate but have smaller or bigger substituents may not cause the conformational change required for catalysis. Post and Ray showed that a unique form of an enzyme is not necessary for effective catalysis of a reaction on the basis of site-directed mutagenesis research. Site-directed mutagenesis means that one amino acid residue in the enzyme is genetically modified to a different amino acid by an enzyme-catalyzed reaction. This would explain why an enzyme-catalyzed reaction only takes place when the substrates bind in a specific order in the case of bimolecular systems, where the binding of the first substrate may cause the conformational shift that exposes the binding site for the second substrate. The induced-fit hypothesis necessitates a flexible active site in contrast to the lock-and-key hypothesis, which assumes a hard active site to accommodate various binding modes and conformational variations in the enzyme. The idea of a flexible active site was first introduced by Pauling who proposed that an enzyme is a flexible template that is most complimentary to substrates at the transition state as opposed to the ground state.

Haldane proposed in 1930 that an enzyme-substrate (E-S) complex needs extra activation energy, and this energy may come from substrate energy on the enzyme. The enzyme conforms to the transition-state structure as the reaction progresses, resulting in the increased binding energy with the transition-state structure. This means that the substrate does not bind efficiently in the ES complex. Rate enhancement is the outcome of this strong binding, also known as transition-state stability. The substrate binds to the active site to create the E/S complex. The active site may consist of only a dozen or so amino acid residues, and only two or three of these may be directly involved in substrate binding and/or catalysis. The protein might also be used to direct the substrate toward the active site.

9.3. MODIFICATIONS

Isocitrate dehydrogenase structural modification research provided proof in favor of the orbital steering theory. The cofactors like organic molecules or metal ions necessary for catalysis for this structure were slightly altered, which resulted in a minute misalignment of the bound cofactors. Because one of the cofactors must react with the substrate during catalysis, misalignment of the cofactor would result in a disrupted reaction trajectory and should have an impact on the enzyme's ability to catalyze. As shown by X-ray crystallographic analyzes of the active isocitrate dehydrogenase complexes, the reaction with the modified cofactors actually led to significant reductions in reaction rate with only minor changes in the orientation of the substrates. Therefore, it seems that even small modifications to the reaction course caused by incorrect orbital alignment might have a significant impact on the catalytic process. Two characteristics like specificity and rate acceleration define enzyme catalysis. These characteristics of an enzyme are caused by moieties found in the active site, specifically amino acid residues and, in the case of some enzymes, cofactors. A cofactor, also known as a coenzyme, is an organic molecule or metal ion that binds to the active site of those enzymes that need cofactors.

The Michaelis complex arises with a diffusional rate constant for catalytically effective enzymes. Transition-state formation, product formation, and product release must all take place within this time window in order to achieve the maximum catalytic rate (McClendon, Zhadin, & Callender, 2005). About 10^{12} times more strongly than it binds the substrate or products, an enzyme binds the transition state structure. Following bond breaking (or making) at the transition state, the interactions that are in the complex that stabilizes the transition-state are no longer present in the complex that contains the products. As a result, the products are poorly bound, which causes them to be released from the active site. Even more significant than the disappearance of binding interactions is the possibility of an alteration in the electronic distribution as bonds are formed and broken. This can result in an attractive interaction between the products or with groups at the active site, which can cause the active site to open and cause the products to be released. An investigation of the transition-state structure for nucleoside hydrolase can explain this process. This study used kinetic isotope effects to create a geometric model of the transition state, and molecular orbital (MO) calculations are then used to establish the molecular electrostatic potential surface of this geometric model. It was discovered that the electrostatic potential surfaces of the transition-state structure and the substrate and products that are attached to the enzymes differ significantly. The enzyme-bound products had some regions dominated by negative charge; this electrostatic repulsion is thought to have destroyed the product affinity for the active site and caused their release.

9.4. INTERACTIONS

A molecule that is a good electron donor may transfer some of its charge to the acceptor when it interacts with a molecule that is a good electron acceptor. A molecular dipole-dipole interaction results in the formation of a charge transfer complex. The difference between the donor's ionization potential and the acceptor's electron affinity determines how energetic this bond could exist. Alkenes, alkynes, and aromatic moieties containing electron-donating substituents are examples of donor groups. Other examples of donor groups include groups having two unbound electrons, such as oxygen, nitrogen, and sulfur moieties. Alkenes, alkynes, aromatic moieties with electron-withdrawing substituents, and weakly acidic protons are examples of molecules with electron-deficient p orbitals in acceptor groups. The carboxylate group of aspartates or the aromatic ring of tyrosine are examples of groups of receptors that can act as electron donors, acceptors, or both. Histidine, tryptophan, and asparagine are examples of groups that can act as both donors and acceptors. The surrounding water molecules orient themselves and are in a higher energy state than when only other water molecules are present in the presence of a nonpolar molecule or part of a molecule.

Because of the higher entropy of the nearby water molecules, the nonpolar group free energy has decreased. According to Jencks, the single mechanism causing noncovalent intermolecular interactions in aqueous solution may be hydrophobic forces. Atoms in nonpolar compounds may temporarily have an asymmetrical distribution of their electron densities, which forms a transient dipole . The temporary dipoles of one molecule cause opposite dipoles in other molecules as atoms from different molecules such as an enzyme and a substrate approach one another. Van der Waals forces, an intermolecular attraction, are the outcome. However, when there is molecular complementarity, numerous atomic interactions result, which can add up to overall enzyme-substrate binding component. These weak universal forces only become important when there is a very close surface contact of the atoms. The presence of several types of bonds/interactions is essential since noncovalent interactions are typically weak. Translational entropy, or the energy associated with the molecules to move is lost after the initial reaction. As a result, the creation of the second interaction has a low entropy loss. Due to this, a number of relatively weak interactions may come together to form a strong interaction. Selectivity in enzyme-substrate interactions can emerge from the involvement of a variety of distinct interactions. The transition state of the reaction is when there are binding interactions at the active site. In order to reduce the free-energy difference between the ground state or intermediate state and the transition state, it is crucial that an enzyme does not bind to the ground state or intermediate states excessively. The substrate is prepared for the process that the enzyme catalyzes via binding interactions between substrate and active site of the enzyme.

9.5. DIASTEREOMERS

Diastereomers differ in their energies, stability, and reactivity, as well as other characteristics When an enzyme is exposed to a racemic mixture of a substrate, the binding energy for the formation of an E-S complex with one enantiomer may be significantly higher than that with the other enantiomer for a variety of reasons, including binding interactions or steric considerations (Fiedler, Leung, Bergman, & Raymond, 2005). The benzyl group at the stereogenic center of phenylalanine, for instance, has two possible orientations after the ammonium and carboxylate substituents have interacted with active-site groups. In the case of the S-isomer, there is a binding pocket, but the benzyl group of the R-isomer points in the opposite direction toward the leucine side chain and causes steric hindrance. Only one E-S complex may develop if the binding energies for the two complexes are considerably different. Only one E-S complex may result in the production of a product due to steric or electronic constraints. Due to the fact that geometric isomers are already diastereomers, enzymes can also exhibit perfect stereospecificity. In general, catalysts stabilize the transition state in relation to the ground state, and the rate acceleration that follows is caused by this decrease in activation energy. Jencks suggested that the ability of enzymes to use binding interactions away from the site of catalysis is the essential characteristic separating enzymes from simple chemical catalysts. By placing substrates in relation to one another and to the catalytic groups at the active site, these binding interactions enable reactions. an.The significance of binding interactions to catalysis was proven by the use of an RNA enzyme. Multiple steps, each with a low activation energy, may be involved because an enzyme has many chances to invoke catalysis, such as by stabilizing transition states, by destabilizing the E-S complex, by destabilizing intermediates, and during product release. This process involves an important change in the protein shape. A reversible reaction equilibrium is unaffected by enzyme catalysis. An enzyme can only speed up the attainment of equilibrium; it cannot speed up the relative concentrations of substrates and products at equilibrium. If an enzyme speeds up the forward reaction, it must also speed up the corresponding backward reaction by the same amount.

9.6. RATE CONSTANT

Although it is impossible to compare first and second-order rate constants, the efficiency of an intramolecular reaction can be expressed in terms of its effective molarity (EM), also known as the effective concentration, which is the concentration of reactant necessary to cause the intermolecular reaction to proceed at the observed rate of the intramolecular reaction. The EM is computed by dividing the intramolecular reaction's first-order rate constant by the equivalent intermolecular reaction's second-order rate constant. The efficiency of intramolecular catalysis varies with structure and can be as high as 10^{16} M for reactive systems.

Holding groups close together can therefore play a substantial role in catalysis, especially when the reactive moieties in an enzyme-substrate complex are aligned correctly for reaction. In order to form covalent bonds with the substrate, some enzymes can use nucleophilic amino acid side chains, such as acidic groups (aspartate or glutamate carboxylates), neutral groups (serine hydroxyl or cysteine thiol), or basic groups (lysine amino, arginine guanidino, or histidine imidazolyl) or cofactors in the active site. The thiol group of cysteine, the hydroxyl group of serine, the imidazole group of histidine, the amino group of lysine, and the carboxylate group of aspartate or glutamate are the most prevalent active-site nucleophiles. A histidine imidazole or a water molecule that has been deprotonated in a base reaction will activate these active-site nucleophiles by deprotonation. The main catalytic benefit of using an active-site residue instead of water is that the former results in a unimolecular reaction, which is entropically preferred over the bimolecular reaction with water. Alkoxides (serine ions) and thiolates are superior nucleophiles to hydroxide ions. Some enzymes may require covalent catalysis to help them bind because they lack a sufficient number of noncovalent binding interactions.

9.7. CATALYSIS

Acid catalysis and/or base catalysis can be important mechanisms for specificity and increasing rate in any reaction where proton transfer takes place (Yang, Liu, Wei, Gu, & Shi, 2022). Consider the enzyme α-chymotrypsin, a serine protease, which uses a serine residue in the active site at a covalent catalytic cleavage of peptide bonds, as an example of acid/base (and covalent) catalysis. Serine should be a weak nucleophile since its nucleophilic group is a hydroxyl. Desolvation, or the removal of water molecules from charged groups at the active site is one method of ground-state instability. A charged group on the substrate would become unstable as a result of being exposed to an environment with a reduced dielectric constant. Desolvation may also reveal a charged group that is water-bonded to the active site, allowing it to contribute more potently through electrostatic catalysis to the stabilization of a charge created at the transition state. The charged groups in the low-dielectric environment effectively maintain charges at the reaction transition state because these electrostatic interactions are stronger there than in water. Different groups on the substrate connect with certain active-site functional groups as the substrate approaches the enzyme, and this reciprocal interaction causes a conformational change in the enzyme active site. By destabilizing the enzyme or by causing appropriate alignment of the active-site groups involved in catalysis, which may be responsible for the start of catalysis, this can cause the enzyme to transition from a low-catalytic form to a high-catalytic one. The majority of enzymes have a portion of active site in a structure that is complementary to the substrate to allow binding on the initial collisional event. The substrate may also undergo deformation, which would result in strain in the substrate. The conformational change need not only occur in the enzyme. It was proposed that the enzyme was elastic and could return to its initial shape when the product was released. This explains how substrates in high-energy conformations can bind to enzymes.

The enzyme α-chymotrypsin has been extensively examined using a variety of methods over a long period of time, making it a great choice for explaining enzymology. It serves as the model for many different types of enzymes, not just proteases. α-chymotrypsin hydrolyzes proteins and peptides through covalent catalysis. Asp, His, and Serine form a catalytic triad that functions as a charge relay system, which is a common mechanism for many proteases. Some researchers think that the connection between Asp and His includes a low-barrier hydrogen bond, while others think that the aspartate anion and the protonated histidine cation interact electrostatically. In addition, a peptidase in aqueous buffers, chymotrypsin also catalyzes the hydrolysis of peptides in n-octane. For instance, N-acetyl-l-histidine methyl ester is 20 times more reactive in n-octane than N-acetyl-l-phenylalanine ethyl ester, despite being just 0.5% as reactive in aqueous buffer.

All animal species have the enzyme acetylcholinesterase, which catalyzes the fast breakdown of the neurotransmitter acetylcholine to stop impulse transmission at cholinergic synapses. The demand for quick neurotransmission termination causes acetylcholinesterase to move closer to a diffusion-controlled reaction in terms of speed. The anionic site, which binds the quaternary ammonium cation of acetylcholine, and the esteratic site, which catalyzes the hydrolysis of the acetyl group, have long been thought to be two significant binding sites in the active site of the enzyme. The active site of other serine hydrolases was thought to resemble the esteratic site. Serine and other organophosphorus nerves disrupt phosphorylate acetylcholinesterase active-site serine irreversibly, causing uncontrollable cholinergic activity that first causes weakness and muscular jerks before leading to full paralysis and muscle atrophy. However, reversible acetylcholinesterase inhibitors cause a mild enhancement of cholinergic activity by facilitating impulse transmission across neuromuscular junctions, which is helpful in treating myasthenia gravis, and by raising acetylcholine levels in the central cholinergic synapses involved in the memory circuit, which explains its benefit in treating Alzheimer's disease. In organic synthesis, chiral catalysts such as esterase and lipases are frequently utilized. An enantioselective catalyst must accept a wide range of substrate configurations and maintain a high level of enantioselectivity for each in order to be effective in synthetic processes. Particularly helpful in this regard are lipases. These catalysts may catalyze the enantioselective hydrolysis and transesterification of a variety of natural and synthetic esters. Guidelines for enantioselectivity have been developed, based on shared structural traits among substrates. Researchers discovered that the slower-reacting enantiomer disrupts the hydrogen bond from the histidine of the catalytic triad to the alcohol oxygen atom, which likely explains the decrease in the rate of hydrolysis of that enantiomer, by comparing crystal structures of enantiomeric transition-state analogs bound to lipase. The organic cofactor known as a heme, which is found in met-myoglobin, is a nonenzymatic protein that is employed to bind and transport oxygen. Two histidine residues that could function as acid/base catalysts for hydrolysis processes are also present in this cavity. The hydrolysis of para-nitrophenyl esters can be catalyzed by this apoprotein, which has esterase activity, at a rate that is 1,900 times faster than that of buffer. This synthetic enzyme is referred to as a semisynthetic enzyme due to its catalytic activity.

9.8. HYDROLYSIS

Hydrolysis of esters has also been studied using catalytic antibodies, sometimes known as abzymes (Hilvert, 2004). Large proteins known as antibodies are made up of two identical heavy chains and two identical light chains that are arranged roughly in the shape of a Y. Controlled levels of ammonia are produced from glutamine at the intended site of action to prevent the toxicity of this molecule. Glutamate is a harmless, inert, and neutral source of ammonia.

An unreactive substance with a structural resemblance to an enzyme substrate or product is referred to as a mechanism-based inactivator. Since they are essential substrates for the enzymes, mechanism-based enzyme inactivators are highly helpful for studying enzyme processes. They were accidentally transformed into reactive products (or intermediates), and these products inactivate the enzyme. However, the processes that transform these substances into their active state use the regular catalytic mechanism. Protein tyrosine phosphatase and protein serine/threonine phosphate residues are hydrolyzed by enzymes belonging to the tyrosine phosphatase and serine/threonine phosphatase superfamilies, respectively. Proteins undergo structural changes as a result of reversible protein phosphorylation, which in turn activates crucial signal transduction pathways involved in cell growth, proliferation, and differentiation. The protein tyrosine phosphatases do not require metal ions or other cofactors for catalysis, but the protein serine/threonine phosphatases are binuclear metallohydrolases, which is a significant distinction in the structure and mechanism of these two superfamilies. The phosphate oxyanions attach to the active-site motif, which is shared by all members of the tyrosine phosphatase superfamily and consists of a cysteine residue and an arginine residue separated by five residues. Each member of this superfamily contains a conserved aspartic acid residue as well. In an associative mechanism, the metal ions would stabilize the transition state to a greater extent than the ground state (two oxyanions), favoring an associative mechanism. As a result, the metal ions stabilize the ground-state structure of a dissociative mechanism (two oxyanions) better than the metaphosphate intermediate (one oxyanion). Activation of the nucleophile (water) is not required due to the high reactivity of a metaphosphate intermediate, but in an associative mechanism this activation is crucial; the crystal structures demonstrate that the nucleophilic water molecule is activated by metal ion interaction. As Lewis acids, the metal ions increase the nucleophilicity by making it more acidic and encouraging the production of hydroxide ions as well as phosphorus electrophilicity.

Molecular oxygen, which undergoes a two-electron reduction to hydrogen peroxide or a four-electron reduction to water, is the most frequent electron acceptor in biological systems. There are no readily apparent redox groups on the side chains of amino acids, which means that in order to remove or add electrons from .a substrate, respectively, an electron acceptor or donor is needed. As a result, coenzymes, or cofactors, also referred to as transition metals or organic molecules, are needed in the active site of redox enzymes in order to facilitate electron transfer. Protons are not exchanged with the solvent during hydride transfer reactions; hence deuterium is not incorporated when these processes are conducted in D_2O. However, the absence of solvent proton exchange in an enzymatic reaction does not always indicate that a hydride transfer mechanism has taken place. An extremely hydrophobic active site could have little to no water in it. Due to exclusion of solvent water molecules, proton transfers can take place in this situation without the need for solvent exchange. Even while an enzyme will catalyze a specific mechanism when its substrate is present, additional mechanisms may appear when the substrate is correctly changed. Unfortunately, natural substrates are frequently ineffective for detecting intermediates in enzyme-catalyzed reactions due to the efficiency of enzyme catalysis. In order to investigate the generation of intermediates, enzyme processes are often examined using modified substrates.

Bibliography

1. Afonso, M. S., De O Silva, A. M., Carvalho, E. B., Rivelli, D. P., Barros, S. B., Rogero, M. M., & Mancini-Filho, J., (2013). Phenolic compounds from rosemary (*Rosmarinus officinalis* L.) attenuate oxidative stress and reduce blood cholesterol concentrations in diet-induced hypercholesterolemic rats. *Nutrition & Metabolism, 10*(1), 19. https://doi.org/10.1186/1743-7075-10-19.
2. Akiyama, T., Sugawara, M., & Arakawa, Y., (2007). Quantum-dot semiconductor optical amplifiers. *Proceedings of the IEEE, 95*(9), 1757–1766. https://doi.org/10.1109/JPROC.2007.900899.
3. Alvarez-Ordóñez, A., & Prieto, M., (2012). Technical and methodological aspects of Fourier transform infrared spectroscopy in food microbiology research. In: *Fourier Transform Infrared Spectroscopy in Food Microbiology* (pp. 1–18). Boston, MA: Springer US. https://doi.org/10.1007/978-1-4614-3813-7_1.
4. Asher, W. E., Pankow, J. F., Erdakos, G. B., & Seinfeld, J. H., (2002). Estimating the vapor pressures of multi-functional oxygen-containing organic compounds using group contribution methods. *Atmospheric Environment, 36*(9), 1483–1498. https://doi.org/10.1016/S1352-2310(01)00564-7.
5. Asmare, M. M., Nitin, N., Yun, S. I., & Mahapatra, R. K., (2022). QSAR and deep learning model for virtual screening of potential inhibitors against inosine 5' monophosphate dehydrogenase (IMPDH) of cryptosporidium parvum. *Journal of Molecular Graphics and Modeling, 111*, 108108. https://doi.org/10.1016/j.jmgm.2021.108108.

6. Baev, A. K., (2012). Types and energies of specific intermolecular interactions in ethers and organic cyclic oxides. In: *Specific Intermolecular Interactions of Organic Compounds* (pp. 63–130). Berlin, Heidelberg: Springer. https://doi.org/10.1007/978-3-642-21622-0_3.

7. Bakke, S. S., Moro, C., Nikolić, N., Hessvik, N. P., Badin, P. M., Lauvhaug, L., & Rustan, A. C., (2012). Palmitic acid follows a different metabolic pathway than oleic acid in human skeletal muscle cells; lower lipolysis rate despite an increased level of adipose triglyceride lipase. *Biochimica et Biophysica Acta (BBA) – Molecular and Cell Biology of Lipids, 1821*(10), 1323–1333. https://doi.org/10.1016/j.bbalip.2012.07.001.

8. Balasubramanian, A., & Ponnuraj, K., (2010). Crystal structure of the first plant urease from jack bean: 83 years of journey from its first crystal to molecular structure. *Journal of Molecular Biology, 400*(3), 274–283. https://doi.org/10.1016/j.jmb.2010.05.009.

9. Bancroft, W. D., & George, A. B., (1930). The reduction of phenol. *Transactions of the American Electrochemical Society, 57*(1), 399. https://doi.org/10.1149/1.3492174.

10. Bent, H. A., (1960). *Distribution of Atomic s Character in Molecules and its Chemical Implications [Research-Article].* https://doi.org/10.1021/ed037p616.

11. Bertsch, W., & Ren, Q., (2000). Chapter 18; the chemical analysis of fire debris for potential accelerants. In: Bogusz, M. J., (ed.), *Handbook of Analytical Separations* (pp. 617–678). Elsevier Science B.V. https://doi.org/10.1016/S1567-7192(00)80073-X.

12. Brattain, W. H., & Garrett, C. G. B., (1955). Experiments on the interface between germanium and an electrolyte. *Bell System Technical Journal, 34*(1), 129–176. https://doi.org/10.1002/j.1538-7305.1955.tb03766.x.

13. Brewer, L., (1968). Bonding and structures of transition metals. *Science, 161*(3837), 115–122. https://doi.org/10.1126/science.161.3837.115.

14. Busch, P., Heinonen, T., & Lahti, P., (2007). Heisenberg's uncertainty principle. *Physics Reports, 452*(6), 155–176. https://doi.org/10.1016/j.physrep.2007.05.006.

15. Cai, X., Ding, S., Li, Z., Zhang, X., Wen, K., Xu, L., & Shen, T., (2021). Simultaneous sintering of low-melting-point Mg with high-melting-point Ti via a novel one-step high-pressure solid-phase sintering strategy. *Journal of Alloys and Compounds, 858*, 158344. https://doi.org/10.1016/j.jallcom.2020.158344.

16. Calixto, J. B., (2019). The role of natural products in modern drug discovery. *Anais Da Academia Brasileira de Ciências, 91*. https://doi.

org/10.1590/0001-3765201920190105.

17. Chattaraj, P. K., & Maiti, B., (2001). Electronic structure principles and atomic shell structure. *Journal of Chemical Education*, *78*(6), 811. https://doi.org/10.1021/ed078p811.
18. Chattopadhyay, P., & Gupta, R. B., (2001). Production of griseofulvin nanoparticles using supercritical CO_2 antisolvent with enhanced mass transfer. *International Journal of Pharmaceutics*, *228*(1), 19–31. https://doi.org/10.1016/S0378-5173(01)00803-1.
19. Cheng, T., Zhao, Y., Li, X., Lin, F., Xu, Y., Zhang, X., & Lai, L., (2007). Computation of octanol–water partition coefficients by guiding an additive model with knowledge. *Journal of Chemical Information and Modeling*, *47*(6), 2140–2148. https://doi.org/10.1021/ci700257y.
20. Chidambaram, R., (1962). Structure of the hydrogen-bonded water molecule in crystals. *The Journal of Chemical Physics*, *36*(9), 2361–2365. https://doi.org/10.1063/1.1732889.
21. Coughtrie, M. W. H., (2016). Function and organization of the human cytosolic sulfotransferase (SULT) family. *Chemico-Biological Interactions*, *259*, 2–7. https://doi.org/10.1016/j.cbi.2016.05.005.
22. Creager, A. N. H., (2021). To test or not to test: Tools, rules, and corporate data in US chemicals regulation. *Science, Technology, & Human Values*, *46*(5), 975–997. https://doi.org/10.1177/01622439211013373.
23. Damalas, C. A., & Eleftherohorinos, I. G., (2011). Pesticide exposure, safety issues, and risk assessment indicators. *International Journal of Environmental Research and Public Health*, *8*(5), 1402–1419. https://doi.org/10.3390/ijerph8051402.
24. Debia, M., & Krishnan, K., (2010). Quantitative property-property relationships for computing occupational exposure limits and vapor hazard ratios of organic solvents. *SAR and QSAR in Environmental Research*, *21*(7, 8), 583–601. https://doi.org/10.1080/1062936X.2010.528241.
25. Denmark, S. E., & Fu, J., (2003). Catalytic enantioselective addition of allylic organometallic reagents to aldehydes and ketones. *Chemical Reviews*, *103*(8), 2763–2794. https://doi.org/10.1021/cr020050h.
26. Dewulf, J., Van, L. H., & Wittmann, G., (2002). Analysis of volatile organic compounds using gas chromatography. *TrAC Trends in Analytical Chemistry*, *21*(9), 637–646. https://doi.org/10.1016/S0165-9936(02)00804-X.
27. Dodabalapur, A., (1997). Organic light emitting diodes. *Solid State Communications*, *102*(2), 259–267. https://doi.org/10.1016/S0038-1098(96)00714-4.

28. Du, Z., Mo, J., Zhang, Y., & Xu, Q., (2014). Benzene, toluene, and xylenes in newly renovated homes and associated health risk in Guangzhou, China. *Building and Environment*, *72*, 75–81. https://doi.org/10.1016/j.buildenv.2013.10.013.
29. Duffy, D., Barrett, M., & Glennon, B., (2013). Novel, calibration-free strategies for supersaturation control in antisolvent crystallization processes. *Crystal Growth & Design*, *13*(8), 3321–3332. https://doi.org/10.1021/cg301673g.
30. Duruz, J. J., Michels, H. J., & Ubbelohde, A. R. J. P., (1997). Molten fatty acid salts as model ionic liquids. I. Thermodynamic and transport parameters of some organic sodium salts. *Proceedings of the Royal Society of London. A. Mathematical and Physical Sciences*, *322*(1550), 281–299. https://doi.org/10.1098/rspa.1971.0068.
31. Dutta, M. J., & Manna, I., (2003). Laser processing of materials. *Sadhana*, *28*(3), 495–562. https://doi.org/10.1007/BF02706446.
32. Elder, D. P., Snodin, D., & Teasdale, A., (2011). Control and analysis of hydrazine, hydrazides, and hydrazones—Genotoxic impurities in active pharmaceutical ingredients (APIs) and drug products. *Journal of Pharmaceutical and Biomedical Analysis*, *54*(5), 900–910. https://doi.org/10.1016/j.jpba.2010.11.007.
33. Fallon, S., (2011). Radiocarbon (14C): Dating and corals. In: Hopley, D., (ed.), *Encyclopedia of Modern Coral Reefs: Structure, Form, and Process* (pp. 829–834). Dordrecht: Springer Netherlands. https://doi.org/10.1007/978-90-481-2639-2_127.
34. Feng, W., & Nansheng, D., (2000). Photochemistry of hydrolytic iron (III) species and photoinduced degradation of organic compounds. A minireview. *Chemosphere*, *41*(8), 1137–1147. https://doi.org/10.1016/S0045-6535(00)00024-2.
35. Fernandez, M., Woo, T. K., Wilmer, C. E., & Snurr, R. Q., (2013). Large-scale quantitative structure–property relationship (QSPR) analysis of methane storage in metal-organic frameworks. *The Journal of Physical Chemistry C*, *117*(15), 7681–7689. https://doi.org/10.1021/jp4006422.
36. Fidaleo, M., & Moresi, M., (2011). Electrodialytic desalting of model concentrated NaCl brines as such or enriched with a non-electrolyte osmotic component. *Journal of Membrane Science*, *367*(1), 220–232. https://doi.org/10.1016/j.memsci.2010.10.069.
37. Fiedler, D., Leung, D. H., Bergman, R. G., & Raymond, K. N., (2005). Selective molecular recognition, C–H bond activation, and catalysis in nanoscale reaction vessels. *Accounts of Chemical Research*, *38*(4), 349–358. https://doi.org/10.1021/ar040152p.

38. Fogassy, E., Nógrádi, M., Kozma, D., Egri, G., Pálovics, E., & Kiss, V., (2006). Optical resolution methods. *Organic & Biomolecular Chemistry, 4*(16), 3011–3030. https://doi.org/10.1039/B603058K.

39. Foster, J. P., & Weinhold, F., (2002). *Natural Hybrid Orbitals.* https://doi.org/10.1021/ja00544a007.

40. Froyen, S., & Harrison, W. A., (1979). Elementary prediction of linear combination of atomic orbitals matrix elements. *Physical Review B, 20*(6), 2420–2422. https://doi.org/10.1103/PhysRevB.20.2420.

41. Gelsomino, A., Petrovičová, B., Tiburtini, S., Magnani, E., & Felici, M., (1997). Multiresidue analysis of pesticides in fruits and vegetables by gel permeation chromatography followed by gas chromatography with electron-capture and mass spectrometric detection. *Journal of Chromatography A, 782*(1), 105–122. https://doi.org/10.1016/S0021-9673(97)00464-0.

42. Georg, R. B., Reynolds, B. C., Frank, M., & Halliday, A. N., (2006). New sample preparation techniques for the determination of Si isotopic compositions using MC-ICPMS. *Chemical Geology, 235*(1), 95–104. https://doi.org/10.1016/j.chemgeo.2006.06.006.

43. Gillespie, R. J., (1963). *The Valence-Shell Electron-Pair Repulsion (VSEPR) Theory of Directed Valency [Research-Article].* https://doi.org/10.1021/ed040p295.

44. Gillespie, R. J., (2002). *Bond Angles and the Spatial Correlation of Electrons.* https://doi.org/10.1021/ja01508a004.

45. Gillman, M. A., & Lichtigfeld, F. J., (1994). Opioid properties of psychotropic analgesic nitrous oxide (laughing gas). *Perspectives in Biology and Medicine, 38*(1), 125–138. https://doi.org/10.1353/pbm.1994.0026.

46. Gorczyca, I., Svane, A., & Christensen, N. E., (1997). Calculated defect levels in GaN and AlN and their pressure coefficients. *Solid State Communications, 101*(10), 747–752. https://doi.org/10.1016/S0038-1098(96)00689-8.

47. Gregory, J. W., Asai, K., Kameda, M., Liu, T., & Sullivan, J. P., (2008). A review of pressure-sensitive paint for high-speed and unsteady aerodynamics. *Proceedings of the Institution of Mechanical Engineers, Part G: Journal of Aerospace Engineering, 222*(2), 249–290. https://doi.org/10.1243/09544100JAERO243.

48. Grozdanov, T. P., & Solov'ev, E. A., (2020). Hidden-crossing explanation of frozen-planet resonances in antiprotonic helium; their positions

and widths. *The European Physical Journal D*, *74*(3), 50. https://doi.org/10.1140/epjd/e2020-100565-0.

49. Guo, H., Peng, Q., Chen, X. K., Gu, Q., Dong, S., Evans, E. W., & Li, F., (2019). High stability and luminescence efficiency in donor–acceptor neutral radicals not following the Aufbau principle. *Nature Materials*, *18*(9), 977–984. https://doi.org/10.1038/s41563-019-0433-1.
50. Henrici-Olivé, G., & Olivé, S., (1979). Molecular interactions and macroscopic properties of polyacrylonitrile and model substances. *Chemistry*, 123–152. Berlin, Heidelberg: Springer. https://doi.org/10.1007/3-540-09442-3_6.
51. Hilvert, D., (2004). Critical analysis of antibody catalysis. In: *Catalytic Antibodies* (pp. 30–71). John Wiley & Sons, Ltd. https://doi.org/10.1002/3527603662.ch2.
52. Hirtz, J., (1985). The gastrointestinal absorption of drugs in man: A review of current concepts and methods of investigation. *British Journal of Clinical Pharmacology*, *19*(S2), 77S–83S. https://doi.org/10.1111/j.1365-2125.1985.tb02746.x.
53. Hoehler, T. M., Som, S. M., & Kiang, N. Y., (2018). Life's requirements. In: Deeg, H. J., & Belmonte, J. A., (eds.), *Handbook of Exoplanets* (pp. 2795–2816). Cham: Springer International Publishing. https://doi.org/10.1007/978-3-319-55333-7_74.
54. House, J. E., & House, K. A., (2016). Chapter 11—Carbon. In: *Descriptive Inorganic Chemistry* (3rd edn., pp. 163–176). Boston: Academic Press. https://doi.org/10.1016/B978-0-12-804697-5.00011-7.
55. Ittel, S. D., Johnson, L. K., & Brookhart, M., (2000). Late-metal catalysts for ethylene homo- and copolymerization. *Chemical Reviews*, *100*(4), 1169–1204. https://doi.org/10.1021/cr9804644.
56. Jacob, C., Giles, G. I., Giles, N. M., & Sies, H., (2003). Sulfur and selenium: The role of oxidation state in protein structure and function. *Angewandte Chemie International Edition*, *42*(39), 4742–4758. https://doi.org/10.1002/anie.200300573.
57. James, A., John, C., Melekamburath, A., Rajeevan, M., & Swathi, R. S., (2022). A journey toward the heaven of chemical fidelity of intermolecular force fields. *WIREs Computational Molecular Science*, *12*(4), e1599. https://doi.org/10.1002/wcms.1599.
58. Kamlet, M. J., Abboud, J. L. M., Abraham, M. H., & Taft, R. W., (2002). *Linear Solvation Energy Relationships. 23. A Comprehensive Collection of the Solvatochromic Parameters.pi.*.alpha., and.Beta., and Some Methods*

for Simplifying the Generalized Solvatochromic Equation. https://doi.org/10.1021/jo00165a018.

59. Kay, A., & Graetzel, M., (2002). *Artificial Photosynthesis. 1. Photosensitization of Titania Solar Cells with Chlorophyll Derivatives and Related Natural Porphyrins.* https://doi.org/10.1021/j100125a029.
60. Kehrer, J. P., (1993). Free radicals as mediators of tissue injury and disease. *Critical Reviews in Toxicology, 23*(1), 21–48. https://doi.org/10.3109/10408449309104073.
61. Kim, J. W., & Koo, K. K., (2012). Crystallization of glycine by drowning-out combined with fines dissolution and cooling process with *in situ* control using focused beam reflectance measurement and attenuated total reflection – Fourier transform infrared spectroscopy. *Crystal Growth & Design, 12*(10), 4927–4934. https://doi.org/10.1021/cg3008574.
62. Kim, K. H., Ho, C. H., Suh, T. G., Prakash, S., & Bunshah, R. F., (1991). Ultrafine aluminum oxide powder prepared by chemical vapor deposition of trimethylaluminum. *Journal of Materials Engineering, 13*(3), 199–205. https://doi.org/10.1007/BF02834178.
63. Kleidon, A., (2020). *Understanding the Earth as a Whole System: From the Gaia Hypothesis to Thermodynamic Optimality and Human Societies.* arXiv. https://doi.org/10.48550/arXiv.2005.09216.
64. Klyachko, A. A., (2009). *The Pauli Exclusion Principle and Beyond.* arXiv. https://doi.org/10.48550/arXiv.0904.2009.
65. Kohler, (1975). The lewis-langmuir theory of valence and the chemical community, 1920-1928. *Historical Studies in the Physical Sciences, 6,* 431–468. https://doi.org/10.2307/27757346.
66. Korpi, A., Järnberg, J., & Pasanen, A. L., (2009). Microbial volatile organic compounds. *Critical Reviews in Toxicology, 39*(2), 139–193. https://doi.org/10.1080/10408440802291497.
67. Kovalenko, S. A., Schanz, R., Hennig, H., & Ernsting, N. P., (2001). Cooling dynamics of an optically excited molecular probe in solution from femtosecond broadband transient absorption spectroscopy. *The Journal of Chemical Physics, 115*(7), 3256–3273. https://doi.org/10.1063/1.1380696.
68. Kravchenko, Yu. P., Liberman, M. A., & Johansson, B., (1996). Exact solution for a hydrogen atom in a magnetic field of arbitrary strength. *Physical Review A, 54*(1), 287–305. https://doi.org/10.1103/PhysRevA.54.287.
69. Kusters, E., & Lauwerys, R., (1990). Biological monitoring of exposure to monochlorobenzene. *International Archives of Occupational and*

Environmental Health, *62*(4), 329–331. https://doi.org/10.1007/BF00640842.

70. Leaver, I. H., & Milligan, B., (1984). Fluorescent whitening agents—A survey (1974-82). *Dyes and Pigments*, *5*(2), 109–144. https://doi.org/10.1016/0143-7208(84)80008-X.
71. Lemmen, C., & Lengauer, T., (2000). Computational methods for the structural alignment of molecules. *Journal of Computer-Aided Molecular Design*, *14*(3), 215–232. https://doi.org/10.1023/A:1008194019144.
72. Levine, B. G., Stone, J. E., & Kohlmeyer, A., (2011). Fast analysis of molecular dynamics trajectories with graphics processing units—Radial distribution function histogramming. *Journal of Computational Physics*, *230*(9), 3556–3569. https://doi.org/10.1016/j.jcp.2011.01.048.
73. Li, A. J., Pal, V. K., & Kannan, K., (2021). A review of environmental occurrence, toxicity, biotransformation, and biomonitoring of volatile organic compounds. *Environmental Chemistry and Ecotoxicology*, *3*, 91–116. https://doi.org/10.1016/j.enceco.2021.01.001.
74. Ligor, T., Górecka, H., & Buszewski, B., (1998). Sorbents for trapping organic pollutants from air. *International Journal of Occupational Safety and Ergonomics*, *4*(2), 153–167. https://doi.org/10.1080/10803548.1998.11076387.
75. Liljenberg, M., Brinck, T., Herschend, B., Rein, T., Rockwell, G., & Svensson, M., (2010). Validation of a computational model for predicting the site for electrophilic substitution in aromatic systems. *The Journal of Organic Chemistry*, *75*(14), 4696–4705. https://doi.org/10.1021/jo100310v.
76. Little, R. D., (2020). A perspective on organic electrochemistry. *The Journal of Organic Chemistry*, *85*(21), 13375–13390. https://doi.org/10.1021/acs.joc.0c01408.
77. Liu, Z., Arnold, R. G., Betterton, E. A., & Smotkin, E., (2001). Reductive dehalogenation of gas-phase chlorinated solvents using a modified fuel cell. *Environmental Science & Technology*, *35*(21), 4320–4326. https://doi.org/10.1021/es001772y.
78. Lu, X., Tao, S., Hu, H., & Dawson, R. W., (2000). Estimation of bioconcentration factors of nonionic organic compounds in fish by molecular connectivity indices and polarity correction factors. *Chemosphere*, *41*(10), 1675–1688. https://doi.org/10.1016/S0045-6535(00)00050-3.
79. Masamune, S., Kaiho, T., & Garvey, D. S., (1982). Aldol methodology: Synthesis of versatile intermediates, 3-hydroxy-2-vinylcarbonyl

compounds. *Journal of the American Chemical Society, 104*(20), 5521–5523. https://doi.org/10.1021/ja00384a060.

80. Matxain, J. M., Piris, M., Mercero, J. M., Lopez, X., & Ugalde, J. M., (2012). Sp3 hybrid orbitals and ionization energies of methane from PNOF5. *Chemical Physics Letters, 531*, 272–274. https://doi.org/10.1016/j.cplett.2012.02.041.

81. McClendon, S., Zhadin, N., & Callender, R., (2005). The approach to the Michaelis complex in lactate dehydrogenase: The substrate binding pathway. *Biophysical Journal, 89*(3), 2024–2032. https://doi.org/10.1529/biophysj.105.062604.

82. McCracken, G., & Stott, P., (2005). Chapter 1—What is nuclear fusion? In: *Fusion* (pp. 1–5). Burlington: Academic Press. https://doi.org/10.1016/B978-012481851-4/50003-2.

83. McDonald, M. A., Salami, H., Harris, P. R., Lagerman, C. E., Yang, X., Bommarius, A. S., & Rousseau, R. W., (2021). Reactive crystallization: A review. *Reaction Chemistry & Engineering, 6*(3), 364–400. https://doi.org/10.1039/D0RE00272K.

84. Miao, M., Wang, X., Brgoch, J., Spera, F., Jackson, M. G., Kresse, G., & Lin, H., (2015). Anionic chemistry of noble gases: Formation of Mg–NG (NG = Xe, Kr, Ar) compounds under pressure. *Journal of the American Chemical Society, 137*(44), 14122–14128. https://doi.org/10.1021/jacs.5b08162.

85. Moeller, K. D., (2018). Using physical organic chemistry to shape the course of electrochemical reactions. *Chemical Reviews, 118*(9), 4817–4833. https://doi.org/10.1021/acs.chemrev.7b00656.

86. Moldoveanu, S. C., (2010). Chapter 17 pyrolysis of carboxylic acids. In: Serban, C. M., (ed.), *Techniques and Instrumentation in Analytical Chemistry* (pp. 471–526). Elsevier. https://doi.org/10.1016/S0167-9244(09)02817-0.

87. Monferran, M. V., Pesce, S. F., Cazenave, J., & Wunderlin, D. A., (2008). Detoxification and antioxidant responses in diverse organs of *Jenynsia* multidentata experimentally exposed to 1,2- and 1,4-dichlorobenzene. *Environmental Toxicology, 23*(2), 184–192. https://doi.org/10.1002/tox.20326.

88. Montero-Campillo, M. M., Lamsabhi, A. M., Mó, O., & Yáñez, M., (2013). Modulating weak intramolecular interactions through the formation of beryllium bonds: Complexes between squaric acid and BeH_2. *Journal of Molecular Modeling, 19*(7), 2759–2766. https://doi.org/10.1007/s00894-012-1603-0.

89. O'Neill, J., Brock, C., Olesen, A. E., Andresen, T., Nilsson, M., & Dickenson, A. H., (2012). Unravelling the mystery of capsaicin: A tool to understand and treat pain. *Pharmacological Reviews*, *64*(4), 939–971. https://doi.org/10.1124/pr.112.006163.
90. Parker, S., (2007). Introduction, history of lasers and laser light production. *British Dental Journal*, *202*(1), 21–31. https://doi.org/10.1038/bdj.2006.113.
91. Pinaud, B. A., Benck, J. D., Seitz, L. C., Forman, A. J., Chen, Z., Deutsch, T. G., & Jaramillo, T. F., (2013). Technical and economic feasibility of centralized facilities for solar hydrogen production via photocatalysis and photoelectrochemistry. *Energy & Environmental Science*, *6*(7), 1983–2002. https://doi.org/10.1039/C3EE40831K.
92. Rendleman, J. A., (1967). Complexes of alkali metals and alkaline-earth metals with carbohydrates. In: Wolfrom, M. L., & Tipson, R. S., (eds.), *Advances in Carbohydrate Chemistry* (Vol. 21, pp. 209–271). Academic Press. https://doi.org/10.1016/S0096-5332(08)60318-5.
93. Rhee, C. H., Metzger, R. M., & Wiygul, F. M., (1982). CNDO/2-FPP atom-in-molecule polarizabilities. *The Journal of Chemical Physics*, *77*(2), 899–915. https://doi.org/10.1063/1.443905.
94. Rieke, F., & Baylor, D. A., (1998). Single-photon detection by rod cells of the retina. *Reviews of Modern Physics*, *70*(3), 1027–1036. https://doi.org/10.1103/RevModPhys.70.1027.
95. Rioux, J. P., & Myers, R. A. M., (1988). Methylene chloride poisoning: A paradigmatic review. *The Journal of Emergency Medicine*, *6*(3), 227–238. https://doi.org/10.1016/0736-4679(88)90330-7.
96. Robinson, D. A., & Woollard, J. McK., (1982). Introduction to organic chemistry. In: *Chemistry for Colleges and Schools* (pp. 485–487). London: Macmillan Education UK. https://doi.org/10.1007/978-1-349-04540-2_25.
97. Rocke, A. J., (2015). It began with a daydream: The 150th anniversary of the Kekulé benzene structure. *Angewandte Chemie International Edition*, *54*(1), 46–50. https://doi.org/10.1002/anie.201408034.
98. Rodriguez, A., (2016). Integrating spray aeration and granular activated carbon for disinfection by-product control in a potable water system. *Electronic Theses and Dissertations*. Retrieved from https://stars.library.ucf.edu/etd/5620.
99. Roman, M., Van, D. L. H., Gutierrez, L., Vanoppen, M., Post, J. W., Wols, B. A., & Verliefde, A. R. D., (2019). Key physicochemical characteristics governing organic micropollutant adsorption and transport in ion-

exchange membranes during reverse electrodialysis. *Desalination*, *468*, 114084. https://doi.org/10.1016/j.desal.2019.114084.

100. Roostaei, M., Soroush, M., Hosseini, S. A., Velayati, A., Alkouh, A., Mahmoudi, M., & Fattahpour, V., (2020). *Comparison of Various Particle Size Distribution Measurement Methods: Role of Particle Shape Descriptors*. Presented at the SPE international conference and exhibition on formation damage control. One Petro. https://doi.org/10.2118/199335-MS.
101. Ruiz, E. J., Sears, D. N., Pines, A., & Jameson, C. J., (2006). Diastereomeric Xe chemical shifts in tethered cryptophane cages. *Journal of the American Chemical Society*, *128*(51), 16980–16988. https://doi.org/10.1021/ja066661z.
102. Simm, G. N., Türtscher, P. L., & Reiher, M., (2020). Systematic microsolvation approach with a cluster-continuum scheme and conformational sampling. *Journal of Computational Chemistry*, *41*(12), 1144–1155. https://doi.org/10.1002/jcc.26161.
103. Smeenk, C., & Ellis, G., (2017). Philosophy of cosmology. In: Zalta, E. N., (ed.), *The Stanford Encyclopedia of Philosophy (Winter 2017)*. Metaphysics Research Lab, Stanford University. Retrieved from: https://plato.stanford.edu/archives/win2017/entries/cosmology/ (accessed on 17 June 2023).
104. Sorokin, A. B., (2013). Phthalocyanine metal complexes in catalysis. *Chemical Reviews*, *113*(10), 8152–8191. https://doi.org/10.1021/cr4000072.
105. Stenberg, M., Stiblert, L., & Nygren, H., (1986). External diffusion in solid-phase immunoassays. *Journal of Theoretical Biology*, *120*(2), 129–140. https://doi.org/10.1016/S0022-5193(86)80169-2.
106. Sushko, I., Novotarskyi, S., Körner, R., Pandey, A. K., Rupp, M., Teetz, W., & Tetko, I. V., (2011). Online chemical modeling environment (OCHEM): Web platform for data storage, model development and publishing of chemical information. *Journal of Computer-Aided Molecular Design*, *25*(6), 533–554. https://doi.org/10.1007/s10822-011-9440-2.
107. Tamarat, Ph., Maali, A., Lounis, B., & Orrit, M., (2000). 10 years of single-molecule spectroscopy. *The Journal of Physical Chemistry A*, *104*(1), 1–16. https://doi.org/10.1021/jp992505l.
108. Tipton, K. F., Boyce, S., O'Sullivan, J., Davey, G. P., & Healy, J., (2004). Monoamine oxidases: Certainties and uncertainties. *Current Medicinal Chemistry*, *11*(15), 1965–1982. https://doi.org/10.2174/0929867043364810.

109. Tong, Z., Benson, G. C., Wang, L. L., & Lu, B. C. Y., (1996). Excess enthalpies of ternary mixtures consisting of a normal alkane, methyl tert-butyl ether, and tert-amyl methyl ether. *Journal of Chemical & Engineering Data*, *41*(4), 865–869. https://doi.org/10.1021/je960078o.

110. Tremsin, A. S., & Vallerga, J. V., (2020). Unique capabilities and applications of microchannel plate (MCP) detectors with Medipix/Timepix readout. *Radiation Measurements*, *130*, 106228. https://doi.org/10.1016/j.radmeas.2019.106228.

111. Turrentine, J. W., (2002). Reversed electrolysis. *The Journal of Physical Chemistry*. (World). https://doi.org/10.1021/j150096a006.

112. Van, D. B. J., & Khomskii, D., (1999). Double exchange via degenerate orbitals. *Physical Review Letters*, *82*(5), 1016–1019. https://doi.org/10.1103/PhysRevLett.82.1016.

113. Veshapidze, G., Nomura, M., Nishide, T., Rajgara, F. A., Shiromaru, H., Achiba, Y., & Kobayashi, N., (2004). Coplanar dissociation of ethylene and benzene by collision with Ar^{8+}. *Journal of Physics B: Atomic, Molecular and Optical Physics*, *37*(14), 2969. https://doi.org/10.1088/0953-4075/37/14/010.

114. Wang, K., Zhang, J., Lou, L., Yang, S., & Chen, Y., (2004). UV or visible light induced photodegradation of AO7 on TiO_2 particles: The influence of inorganic anions. *Journal of Photochemistry and Photobiology A: Chemistry*, *165*(1), 201–207. https://doi.org/10.1016/j.jphotochem.2004.03.025.

115. Wang, X., Zhou, H., Arnott, W. P., Meyer, M. E., Taylor, S., Firouzkouhi, H., & Watson, J. G., (2020). Evaluation of gas and particle sensors for detecting spacecraft-relevant fire emissions. *Fire Safety Journal*, *113*, 102977. https://doi.org/10.1016/j.firesaf.2020.102977.

116. Weingärtner, H., (2008). Understanding ionic liquids at the molecular level: Facts, problems, and controversies. *Angewandte Chemie International Edition*, *47*(4), 654–670. https://doi.org/10.1002/anie.200604951.

117. Weisel, C. P., & Jo, W. K., (1996). Ingestion, inhalation, and dermal exposures to chloroform and trichloroethene from tap water. *Environmental Health Perspectives*, *104*(1), 48–51. https://doi.org/10.1289/ehp.9610448.

118. Withey, R. J., & Hall, J. W., (1975). The joint toxic action of perchloroethylene with benzene or toluene in rats. *Toxicology*, *4*(1), 5–15. https://doi.org/10.1016/0300-483X(75)90017-7.

119. Woutersen, R. A., Appelman, L. M., Wilmer, J. W. G. M., Falke, H. E., & Feron, V. J., (1987). Subchronic (13-week) inhalation toxicity study of formaldehyde in rats. *Journal of Applied Toxicology*, *7*(1), 43–49. https://

doi.org/10.1002/jat.2550070108.

120. Yang, X., Liu, Z., Wei, G., Gu, Y., & Shi, H., (2022). A critical assessment of the roles of water molecules and solvated ions in acid-base-catalyzed reactions at solid-water interfaces. *Chinese Journal of Catalysis*, *43*(8), 1964–1990. https://doi.org/10.1016/S1872-2067(21)64032-9.
121. Yu, L., (2001). Amorphous pharmaceutical solids: Preparation, characterization and stabilization. *Advanced Drug Delivery Reviews*, *48*(1), 27–42. https://doi.org/10.1016/S0169-409X(01)00098-9.
122. Zeeman, M., & Gilford, J., (1993). Ecological hazard evaluation and risk assessment under EPA's toxic substances control act (TSCA): An introduction. *Environmental Toxicology and Risk Assessment*, 7–21. ASTM International. https://doi.org/10.1520/STP19230S.
123. Zhang, T., Szilágyi, B., Gong, J., & Nagy, Z. K., (2019). Thermodynamic polymorph selection in enantiotropic systems using supersaturation-controlled batch and semibatch cooling crystallization. *Crystal Growth & Design*, *19*(11), 6715–6726. https://doi.org/10.1021/acs.cgd.9b01076.
124. Zhao, X., Huang, W., Song, D., Lin, R., Huang, H., Huang, J., & Ye, G., (2020). The hydrogen transfer reaction between the substance of triplet state thioxanthone and alkane with sp3 hybridization hydrogen. *Journal of Molecular Modeling*, *26*(3), 56. https://doi.org/10.1007/s00894-020-4300-4.
125. Zhao, Y., Bao, Y., Wang, J., & Rohani, S., (2012). *In situ* focused beam reflectance measurement (FBRM), attenuated total reflectance Fourier transform infrared (ATR-FTIR) and Raman characterization of the polymorphic transformation of carbamazepine. *Pharmaceutics*, *4*(1), 164–178. https://doi.org/10.3390/pharmaceutics4010164.
126. Zhao, Y., Song, Y., & Duan, L., (2022). Study on the effect of water flux in osmotic microbial fuel cells on membrane water content and resistance. *Water*, *14*(6), 848. https://doi.org/10.3390/w14060848.
127. Zhou, M., Dai, Q., Lei, L., Ma, C., & Wang, D., (2005). Long life modified lead dioxide anode for organic wastewater treatment: Electrochemical characteristics and degradation mechanism. *Environmental Science & Technology*, *39*(1), 363–370. https://doi.org/10.1021/es049313a.
128. Zucco, G. M., & Doty, R. L., (2022). Multiple chemical sensitivity. *Brain Sciences*, *12*(1), 46. https://doi.org/10.3390/brainsci12010046.
129. Zweers, T., Skov, P., Valbjørn, O., & Mølhave, L., (1990). The effect of ventilation and air pollution on perceived indoor air quality in five town halls. *Energy and Buildings*, *14*(3), 175–181. https://doi.org/10.1016/0378-7788(90)90038-K.

INDEX

B

C

F

G

H

O

P

Q

R

S

T

U

V

W